Heat Exchangers

*Edited by Laura Castro Gómez,
Víctor Manuel Velázquez Flores
and Miriam Navarrete Procopio*

Published in London, United Kingdom

IntechOpen

Supporting open minds since 2005

Heat Exchangers
http://dx.doi.org/10.5772/intechopen.95656
Edited by Laura Castro Gómez, Víctor Manuel Velázquez Flores and Miriam Navarrete Procopio

Contributors
Abdelhanine A. Benallou, Thanh Nhan Phan, Van Hung Tran, Prashant B. Dehankar, I Made Arsana, Ruri Agung Wahyuono, Huihe Qiu, Yinchuang Yang, Shahin Kharaji, Khadijah Lawal, Haruna Jibril, Muhammad Ahmad Jamil, Talha S. Goraya, Haseeb Yaqoob, Kim Choon Ng, Muhammad Wakil Shahzad, Syed M. Zubair, Marip Kum Ja, Qian Chen, Muhammad Burhan, Doskhan Ybyraiymkul, Raid Alrowais, Nguyen Minh Phu, Ngo Thien Tu

First published in London, United Kingdom, 2022 by IntechOpen
IntechOpen is the global imprint of INTECHOPEN LIMITED, registered in England and Wales, registration number: 11086078, 5 Princes Gate Court, London, SW7 2QJ, United Kingdom
Printed in Croatia

British Library Cataloguing-in-Publication Data
A catalogue record for this book is available from the British Library

Additional hard and PDF copies can be obtained from orders@intechopen.com

Heat Exchangers
Edited by Laura Castro Gómez, Víctor Manuel Velázquez Flores and Miriam Navarrete Procopio
p. cm.
Print ISBN 978-1-83969-791-3
Online ISBN 978-1-83969-792-0
eBook (PDF) ISBN 978-1-83969-793-7

For EU product safety concerns:
IN TECH d.o.o., Prolaz Marije Krucifikse Kozulić 3, 51000 Rijeka, Croatia, info@intechopen.com or visit our website at intechopen.com.

Meet the editors

Laura Castro Gómez obtained a Ph.D. in Engineering and Applied Sciences at the Center of Research in Engineering and Applied Science (CIICAp). She is a professor and researcher in mechanical engineering at Morelos State University (UAEM), Mexico. She teaches thermodynamics, fluids mechanics, and heat transfer, among other subjects. She also has expertise in turbomachinery, fluid flows, and heat exchangers. She has published sixteen papers in scientific journals and four book chapters. She has also co-edited two books. Dr. Castro is a reviewer for journals such as the *International Journal Of Energy Research* and *Heat and Mass Transfer*. She is a member of the board of directors of the Mexican Society of Mechanical Engineering, A.C. (SOMIM).

Miriam Navarrete Procopio is a Professor of Chemical Engineering, Morelos State University (UAEM), Mexico. She obtained a Ph.D. in Engineering and Applied Sciences at the Center of Research in Engineering and Applied Science (CIICAp). Dr. Navarrete leads chemical engineering laboratories where she teaches distillation, absorption, desorption, adsorption columns, heat exchangers, and chemical reactors design. She has been conducting exergetic studies of several processes at both the industrial and laboratory scales.

Víctor Velázquez Flores is a Professor of Chemical Engineering, Morelos State University (UAEM), Mexico. He received a BChE and Ph.D. in Applied Chemical Engineering from the same university. Dr. Velázquez has taught thermodynamics, applied thermodynamics, applied fluids mechanics, mass and energy balance, and heat transfer. He has published articles in refereed and indexed journals and given workshops for the training of students and teachers. He has also directed theses and served as a dissertation advisor.

Contents

Preface

Heat exchangers are devices that exchange thermal energy between two or more fluids. They are used in both cooling and heating applications to increase or decrease the temperature of a fluid or develop a phase change. Therefore, thermal energy storage, properties of the fluids involved, and construction materials are relevant parameters for the design of heat exchangers (see **Figure 1**).

Several studies on heat exchangers are focused on heat recovery, energy exchange with novel fluids, and redesign and optimization of geometry.

This book begins by discussing energy-saving and heat recovery as essential factors in heat transfer processes. It presents thermoeconomic analyses that include thermodynamic and economic parameters. It also examines novel fluids and applications in heat transfer. The selection of working fluids is relevant to determine the operating conditions in heat exchangers. Electrically conductive fluids in the presence of magnetic and electric fields are studied by magnetohydrodynamics. A chapter on the numerical analysis of heat transfer by natural convection in a concentric annulus is presented, considering ramping temperature and ramped motion of the boundaries. The book also reviews the state-of-the-art cooling systems in electronic equipment. In addition, it presents some suggested techniques for the design and analysis of heat exchangers for use in heating, cooling, humidification, and dehumidification. The book ends with a discussion of the design and optimization of heat exchangers, including exchangers with phase change that have application in refrigeration systems. Topics of optimization of heat exchangers include operating cost, pressure drops, heat transfer area, and maximum effectiveness.

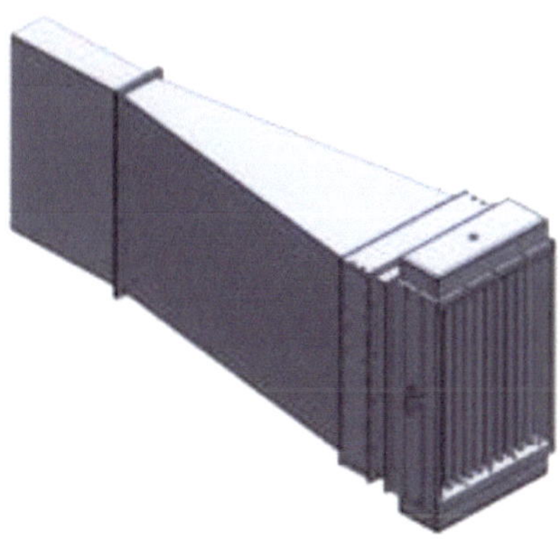

Figure 1.
Heat exchanger air-water.

The editor would like to sincerely thank her co-editors, Miriam Navarrete and Víctor Velázquez, for their valuable support. Without them, this book would not have been possible.

Laura Castro Gómez
Centro de Investigación en Ingeniería y Ciencias Aplicadas,
Universidad Autónoma del Estado de Morelos,
Cuernavaca, Morelos, México

Víctor Manuel Velázquez Flores and Miriam Navarrete Procopio
Universidad Autónoma del Estado de Morelos,
Cuernavaca, Morelos, México

Section 1

Heat Recovery

Chapter 1

Exergoeconomic and Normalized Sensitivity Analysis of Plate Heat Exchangers: A Theoretical Framework with Application

Muhammad Ahmad Jamil, Talha S. Goraya, Haseeb Yaqoob, Kim Choon Ng, Muhammad Wakil Shahzad and Syed M. Zubair

Abstract

Heat exchangers are the mainstay of thermal systems and have been extensively used in desalination systems, heating, cooling units, power plants, and energy recovery systems. This chapter demonstrates a robust theoretical framework for heat exchangers investigation based on two advanced tools, i.e., exergoeconomic analysis and Normalized Sensitivity Analysis. The former is applied as a mutual application of economic and thermodynamic analyses, which is much more impactful than the conventional thermodynamic and economic analyses. This is because it allows the investigation of combinatory effects of thermodynamic and fiscal parameters which are not achieved with the conventional methods. Similarly, the Normalized Sensitivity Analysis allows a one-on-one comparison of the sensitivity of output parameters to the input parameters with entirely different magnitudes on a common platform. This rationale comparison is obtained by normalizing the sensitivity coefficients by their nominal values, which is not possible with the conventional sensitivity analyses. An experimentally validated example of a plate heat exchanger is used to demonstrate the application of the proposed framework from a desalination system.

Keywords: exergoeconomic analysis, normalized sensitivity analysis, heat exchangers, theoretical framework

1. Introduction

Heat exchangers are an essential component of thermal systems and increase system efficiency by recovering heat from the waste streams [1]. Heat exchangers play a vital role in several applications i.e., waste heat recovery, thermal desalination units, power plants, air conditioning, refrigeration, manufacturing industry, food, chemical, and process industries, etc. The water purification industry that fulfills ~40% of water demand worldwide is based on thermal-based desalination systems [2]. These systems include mechanical/thermal vapor compression (TVC/MVC) systems, adsorption systems, multi-effect desalination (MED), and multistage flash (MSF) [3]. These systems are mostly used due to their high operational reliability, ability to use low-grade energy, low pre-and-post treatment requirement, and capability to treat harsh feeds [4]. Thermal-based desalination systems operate at high brine temperature, and several pieces of research have been carried to improve their thermal and economic performance [5]. One of the major improvements in this regard is energy recovery by using a preheater. The additional component recovers heat from the waste stream i.e., brine, and preheat the intake stream which reduces thermal losses, decreases the evaporator loads, area, and investment [6].

Plate heat exchangers (PHXs) are widely used for heat recovery in thermal-based desalination units as a preheater. The plate heat exchanger offers many benefits including narrow temperature control ($\Delta T \leq 5°C$), easy maintenance and cleaning, margin to accommodate different loads, and high operational reliability [7]. Furthermore, it is significant to indicate that PHXs as preheaters have rarely been examined in thermal-based desalination units from an optimized cost design and analysis viewpoint [8]. Rather, the conducted studies either are restricted to preliminary sizing [9] or heat exchanger design is missing [10]. In conventional studies, the heat transfer area is calculated by the temperature-based heat transfer coefficient correlation offered by Dessouky et al. [11]. However, this method gives a fast estimation of heat transfer area, but the accuracy and reliability of the method are doubtful. This is because, in the heat exchangers, the heat transfer coefficient is the function of different parameters such as pressure, temperature, thermophysical properties, flow characteristics, and geometric parameter [12].

For example, in many previous studies, the plate chevron angle (β) is reported as the most influential geometrical variable of PHXs from the thermal–hydraulic performance viewpoint [13]. Likewise, the heat duty, thermophysical properties, and flow rates also have a remarkable impact on PHXs performance [14]. Some recent optimization studies highlighted the importance of various other process and geometric variables that significantly affect the PHXs performance [15]. For instance, the most critical and influential parameters that have been reported are dimensions of chevron corrugation, number of passes, number of plates, type of plate, and channel flow type (parallel, counter, mixed, etc.) [16].

As it appears from the above literature review that there is a requirement for a laborious optimum cost design and detailed investigation of the preheaters for the thermal-based water treatment systems. In this regard, Jamil et al. [17] moderately addressed the issues and conducted a detailed thermal–hydraulic analysis but have deficiencies in the economic analysis viewpoint. This book chapter is focused on the combinatory effect of thermal, hydraulic, and economic analysis. Furthermore, normalized sensitivity analysis and exergoeconomic analysis are also conducted. This chapter will discuss the sections as follow (a) exergoeconomic analysis methodology, (b) normalized sensitivity analysis methodology, (c) experimentally validation of the numerical model, (d) normalized sensitivity analysis in term of NSC and RC, and (e) exergoeconomic analysis. The normalized sensitivity and exergoeconomic analysis are conducted for a preheater (PHX) of a single evaporator based MVC desalination system as a case study.

2. Exergoeconomic analysis methodology

2.1 Heat exchanger configuration

Figure 1 represents the schematic diagram of the current considered system. The system includes PHXs and two centrifugal pumps to maintain the desire flow

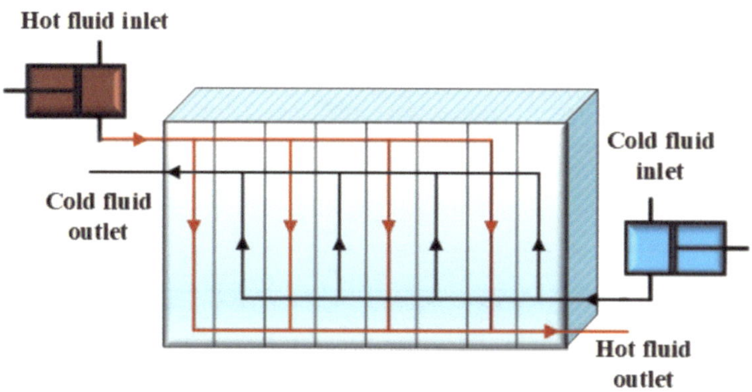

Figure 1.
Plate heat exchanger configuration for current case study.

Parameter		Value
Mass flow rate	Seawater, $\dot{m}_{SW}$ (kg/s)	13
	Brine, $\dot{m}_B$ (kg/s)	13
Temperature of seawater	Inlet, $T_{SW,i}$ (°C)	21
	Outlet, $T_{SW,o}$ (°C)	57
Temperature of Brine	Inlet, $T_{B,i}$ (°C)	63
	Outlet, $T_{B,o}$ (°C)	23
Salinity	Sea water, S_{SW} (g/kg)	40
	Brine, S_B (g/kg)	80

Table 1.
Input operation variables for the current case study [18].

rates and overcome the pressure losses. The PHXs are used as a preheater in single evaporator based MVC water treatment system [18] to preheat the intake seawater using hot brine water. The operational variables i.e., mass flow rates, salinity, the temperature of hot and cold streams are extracted from our recent studies, as mentioned in **Table 1** [18].

2.2 Thermal–hydraulic analysis model

The thermo hydraulic design of the PHXs presented previous study [17] is used for the calculation of different parameters such as flow rates, temperature, area, pressure drop, heat duty, local and global heat transfer coefficient, etc. In the thermal investigation, Nusselt number (Nu) is one of the most important parameters and can be calculated using a correlation (Eq. (1)) which is primarily dependent on the Reynold number (Re) and Prandtl number (Pr) [19].

$$Nu = C_h \, \mathrm{Re}^{\,n} \ \mathrm{Pr}^{0.333} \left(\frac{\mu}{\mu_w}\right)^{0.17} \tag{1}$$

Where the value of C_h and n with different Reynold number and Chevron angle is given in [19]. The governing equations for the calculation of a detailed thermal

Variables	Units	Formula
Reynold number	—	$Re = v_{chl} \times D_{hyd}/\mu$
Mass velocity per channel	kg/m^{2}s	$v_{chl} = \dot{m}/N_{cpp} \times A_{chl}$
Number of channels per pass	—	$N_{cpp} = N_{tb}–1/2 \times N_P$
Single-channel flow area	m^2	$A_{chl} = L_w \times B$
Mean channel flow gap	—	$B=PP–t_{plate}$
Plate pitch	m	$PP = L_c/N_{tb}$
Hydraulic diameter	m	$D_{hyd} = 2\times B/EF$
Projected plate area	m^2	$A_p = (L_v–D_p) \times L_w$
Enlargement factor	—	$A_{sp} = EF \times A_p$
Effective area	m^2	$A = A_{sp} \times N_e$
Effective number of plates	—	$N_e = N_{tb}–2$
Local heat transfer coefficient	kW/m^{2}K	$h = Nu\times k/D_{hyd}$
Overall clean heat transfer coefficient	kW/m^{2}K	$\frac{1}{U_{cl}} = \frac{1}{h_c} + \frac{t_{plate}}{k_{plate}} + \frac{1}{h_h}$
Overall heat transfer coefficient	kW/m^{2}K	$\frac{1}{U_{fo}} = \frac{1}{U_{cl}} + R_{fo,\text{total}}$
Factor of Cleanliness	—	$FOC = U_{fo}/U_{cl}$
Over surface design	%	$OSD = (U_c + R_{fo,\text{total}})\times 100$
Heat duty	kW	$\dot{Q} = A_e \times U \times \Delta T_{LMTD}$

Table 2.
Thermal design equations of PHXs [19].

model are summarized in **Table 2**. While the implementation and selection of correlation are discussed and summarized in [17].

The hydraulic analysis includes the investigation of pumping power and total pressure drop, which is dependent on various pressure losses i.e., ports losses, manifolds losses, and channels losses as shown below [13, 19].

$$\Delta P_{tot} = \Delta P_{chl} + \Delta P_{po} + \Delta P_{man} \tag{2}$$

The pumping power can be calculated as.

$$P_{power} = \frac{\dot{m}\,\Delta P_{tot}}{\eta_p\,\rho} \tag{3}$$

The governing equation of the remaining hydraulic model is summarized in **Table 3**.

2.3 Exergy and exergoeconomic analysis

For the heat exchanger analysis, exergy analysis is a significant and reliable technique because it includes the exergy destruction calculation [20]. The exergy analysis measures overall performance and concurrently responsible for the changes in temperature and pressure. The exergy destruction calculations estimate the performance index of the analysis [21]. For the analysis, the flow exergy is determined at boundaries (inlet and outlet) of pumps and heat exchangers based on their operational

Variables	Units	Formula
Pressure drops in the channel	kPa	$\Delta P_{chl} = 4 \times ff \times \frac{L_e \times N_p}{D_{hyd}} \times \frac{\nu_{chl}^2}{2 \times \rho} \times \left(\frac{\mu}{\mu_w}\right)^{-0.17}$
Pressure drops in ports	kPa	$\Delta P_{po} = 1.4 \times N_P \times \frac{\nu_p^2}{2 \times \rho}$
Portside mass velocity	kg/m^2s	$\nu_p = \frac{\dot{m}}{\pi \times \left(\frac{D_p^2}{4}\right)}$
Pressure drop in manifold	kPa	$\Delta P_{man} = 1.5 \times \left(\frac{V^2}{2 \times v_s}\right)$
Friction factor	—	$ff = \frac{K_P}{\mathrm{Re}^m}$

Table 3.
Hydraulic design equations of PHXs [13, 19].

parameters such as mass flow rates, temperature, pressure, and salinity, as given in Eq. (4). After that, Eq. (6) is solved for all the components to get the exergy destruction. In the present study, the seawater database is used for the calculation of specific flow exergy $\overline{EX}$ and thermophysical properties [22].

$$\overline{EX} = \left[\left(h' - h_0'\right) - T_0(s - s_0)\right] + \overline{EX}_{che} \tag{4}$$

$$\dot{E} = \dot{m} \times \overline{EX} \tag{5}$$

$$\dot{E}_D = \dot{E}_i - \dot{E}_o \tag{6}$$

For the heat exchanger, the economic investigation is depending on the capital/purchasing investment (CI) and operational/running cost (OC) [23]. However, for the large component of the system, such as power plants and desalination units, the product cost is more important than purely capital investment and operational cost [24] because, in these systems, the performance of HX is primarily dependent upon the plant process variables. Therefore, the HX is analyzed and designed to meet the plant requirement [6, 18] instead of optimum HX performance.

The total cost of the heat exchanger is the sum of the capital investment (CI) and operational cost (OC) as given below [25].

$$C_{tot} = CI + OC \tag{7}$$

The capital investment (CI) is the initial amount required to purchase equipment based on time and location of analysis. The finest method to calculate the capital investment to use the experimental correlations purposed by researchers and vendors after extensive study and survey. In the current study, the capital investment of the pump and heat exchanger is calculated using the most common and reliable correlations presented in [26, 27].

The capital investment correlations used for the heat exchanger are generally dependent upon the heat transfer area as [28].

$$CI_{PHX}^{\$} = 1000 \times \left(12.86 + A^{0.8}\right) \times IF \tag{8}$$

After that, an installation factor (IF) range from 1.5 to 2.0 is used to predict accurately the monetary of the equipment at the utility. In contrast, the capital investment of the pump is calculated as [27].

$$CI_P^{\$} = 13.92 \times \dot{m} \times \Delta P^{0.55} \times \left(\eta_p / \left(1 - \eta_p\right)\right)^{1.05} \tag{9}$$

A detailed discussion regarding the capital investment correlation is given in the reference study [29]. Furthermore, the constant in the correlation is varying with material selection and the applicability range. The empirical correlations are developed a long time ago based on the fiscal policy of that era. Therefore, all the above correlations need a slight correction to accurately estimate the capital investment in the current time. In this aspect, the cost index factor (C_{index}) is commonly used. The C_{index} is calculated by using Eq. (10) in which the chemical engineering plant cost index (CEPCI) is used for the original year and the present year as given as [30, 31].

$$C_{index} = \frac{CEPCI_{current}}{CEPCI_{reference}} \tag{10}$$

$$CI^{\$}_{current} = C_{index} \times CI^{\$}_{reference} \tag{11}$$

In the present analysis, the C_{index} 1.7 is used based on their CEPCI 390 [32] and CEPCI 650 for the year of 1990 and 2020 [33] respectively. However, the importance of the Cost index is analyzed from different ranges in the result and discussion section. Likewise, the operation cost (OC) is calculated using Eq. (12). The OC is primarily dependent on the pumping power, P_{Power} (kW), yearly current cost, C_y (\$/y), the unit cost of electricity, C_{ele} (\$/kWh), inflation rate, i (%), operating hours, Φ (h/y), and component life, n_y (year).

$$OC = \sum_{j=1}^{n_y} \frac{C_y}{(1+i)^{\;j}} \tag{12}$$

$$C_o = P_{power} \times C_{ele} \times \Phi \tag{13}$$

$$P_{power} = \frac{1}{\eta_p}\left(\frac{\dot{m}_{SW} \times \Delta P_{SW}}{\rho_{SW}} + \frac{\dot{m}_B \times \Delta P_B}{\rho_B}\right) \tag{14}$$

Whereas, the values operating hours Φ = 7000 h/y, component life n_y = 10 years, unit cost of electricity C_{ele} = 0.09 (\$/KWh) and efficiency of pump η_p = 78% [25] are used in current analysis.

The output cost of the hot stream can be calculated by implementing the general cost approach [18]. For this purpose, the pre-calculated capital investment is converted into the yearly capital investment rate $\dot{\Gamma}$ (\$/y) by using the capital recovery factor (r) [6].

$$r = \frac{i \times (1+i)^{n_y}}{(1+i)^{n_y} - 1} \tag{15}$$

$$\dot{\Gamma} = r \times CI \tag{16}$$

After that, the annual rate is transferred into the fixed cost rate ς (\$/s) through the plant availability factor (Φ).

$$\varsigma = \frac{\dot{\Gamma}}{3600 \times \Phi} \tag{17}$$

After determining the cost flow rate, the cost balance takes the form mentioned below.

$$C_o = \Sigma C_i + \varsigma \tag{18}$$

Whereas the ς is the component cost rate, C_i is the cost of the inlet stream and C_o is the product cost of the outlet stream. The cost balance (refers to Eq. (18)) is re-arranged for the cost balance of the heat exchanger and pump as.

$$C_o = C_i + C_{ele} \times \dot{W}_P + \varsigma_P \tag{19}$$

$$C_{c,o} = C_{c,i} + C_{h,i} - C_{h,o} + \varsigma_{PHX} \tag{20}$$

The cost of the inlet stream is varying from case to case. For the current case study, the inlet cost of the seawater is chosen from the study. It is important to mention that the equipment with various outputs such as RO trains, HXs, flashing stages, evaporation effects, etc.,) need an additional equation for the result. For instance, for the component with "k" outputs, a "k-1" number of additional equations are required. The cost balance of the plate heat exchanger (PHXs) can be solved by using the supplementary equation (Eq. (21)). The equivalency of the average inlet cot and outlet cost of streams depends on these additional Equations [29].

$$\frac{C_{B,i}}{E_{B,i}} - \frac{C_{B,o}}{E_{B,o}} = 0 \tag{21}$$

3. Normalized sensitivity analysis methodology

The sensitivity analysis is an important tool to examine the behavior of output performance parameters against the different input variables [34]. Sensitivity analysis is a significant tool to identify the influential and critical performance parameters and highlights the design improvements for future research. For this purpose, calculus-based (partial derivative-based) sensitivity analysis is one of the most trustworthy and widely used methods. In this approach, all the independent parameters sum up their nominal values and uncertainty as given below [35].

$$X = \overline{X} \pm \widehat{U}_X \tag{22}$$

where $\overline{X}$ and $\pm\widehat{U}_X$ represents the nominal value and the uncertainty about the nominal value, respectively. The uncertainty in the output performance parameter Y(X) because of the uncertainty of variable X is given below [35].

$$\widehat{U}_Y = \frac{dY}{dX}\widehat{U}_X \tag{23}$$

The total uncertainty for the multi-variable function is given as.

$$\widehat{U}_Y = \left[\sum_{j=1}^{N}\left(\frac{\partial Y}{\partial X_j}\widehat{U}_{Xj}\right)^2\right]^{1/2} \tag{24}$$

The partial derivative parameter in the total uncertainty equation denotes the sensitivity coefficient (SC) of the selected output parameter. These SC are converted into modified forms knowns as the Normalized Sensitivity Coefficient (NSC) by regulating the uncertainty in the outlet variable Y and input variable X by their corresponding nominal value ($\overline{X}$). The NSC provides a comparison of all the input variables with significantly different magnitude based on their critical impact on the desired performance parameter [36]. The NSC can be written mathematically as [35].

$$\frac{\widehat{U}_Y}{\overline{Y}} = \left[\sum_{j=1}^{N} \overbrace{\left(\frac{\partial Y}{\partial X_j} \frac{\overline{X}_j}{\overline{Y}} \right)^2}^{NSC} \overbrace{\left(\frac{\widehat{U}_{Xj}}{\overline{X}_j} \right)^2}^{N\widehat{U}_{Xj}} \right]^{1/2} \tag{25}$$

Where NU denotes the normalized uncertainty, and NSC denotes the normalized sensitivity coefficient. Thus, the Eq. (25) can be written for the selected output performance parameters in term of NSC as follow.

$$\frac{\widehat{U}_{h_c}}{\overline{h}_c} = \left[\begin{array}{l} \left(\frac{\partial h_c}{\partial \dot{m}_c} \frac{\overline{\dot{m}}_c}{\overline{h}_c} \right)^2 \left(\frac{\widehat{U}_{\dot{m}_c}}{\overline{\dot{m}}_c} \right)^2 + \left(\frac{\partial h_c}{\partial \dot{m}_h} \frac{\overline{\dot{m}}_h}{\overline{h}_c} \right)^2 \left(\frac{\widehat{U}_{\dot{m}_h}}{\overline{\dot{m}}_h} \right)^2 + \left(\frac{\partial h_c}{\partial T_{c,i}} \frac{\overline{T}_{c,i}}{\overline{h}_c} \right)^2 \left(\frac{\widehat{U}_{T_{c,i}}}{\overline{T}_{c,i}} \right)^2 \\ + \left(\frac{\partial h_c}{\partial T_{h,i}} \frac{\overline{T}_{h,i}}{\overline{h}_c} \right)^2 \left(\frac{\widehat{U}_{T_{h,i}}}{\overline{T}_{h,i}} \right)^2 + \left(\frac{\partial h_c}{\partial S_c} \frac{\overline{S}_c}{\overline{h}_c} \right)^2 \left(\frac{\widehat{U}_{S_c}}{\overline{S}_c} \right)^2 + \left(\frac{\partial h_c}{\partial S_h} \frac{\overline{S}_h}{\overline{h}_c} \right)^2 \left(\frac{\widehat{U}_{S_h}}{\overline{S}_h} \right)^2 \end{array} \right]^{1/2} \tag{26}$$

$$\frac{\widehat{U}_{\Delta P_c}}{\overline{\Delta P}_c} = \left[\begin{array}{l} \left(\frac{\partial \Delta P_c}{\partial \dot{m}_c} \frac{\overline{\dot{m}}_c}{\overline{\Delta P}_c} \right)^2 \left(\frac{\widehat{U}_{\dot{m}_c}}{\overline{\dot{m}}_c} \right)^2 + \left(\frac{\partial \Delta P_c}{\partial \dot{m}_h} \frac{\overline{\dot{m}}_h}{\overline{\Delta P}_c} \right)^2 \left(\frac{\widehat{U}_{\dot{m}_h}}{\overline{\dot{m}}_h} \right)^2 + \left(\frac{\partial \Delta P_c}{\partial T_{c,i}} \frac{\overline{T}_{c,i}}{\overline{\Delta P}_c} \right)^2 \left(\frac{\widehat{U}_{T_{c,i}}}{\overline{T}_{c,i}} \right)^2 \\ + \left(\frac{\partial \Delta P_c}{\partial T_{h,i}} \frac{\overline{T}_{h,i}}{\overline{\Delta P}_c} \right)^2 \left(\frac{\widehat{U}_{T_{h,i}}}{\overline{T}_{h,i}} \right)^2 + \left(\frac{\partial \Delta P_c}{\partial S_c} \frac{\overline{S}_c}{\overline{\Delta P}_c} \right)^2 \left(\frac{\widehat{U}_{S_c}}{\overline{S}_c} \right)^2 + \left(\frac{\partial \Delta P_c}{\partial S_h} \frac{\overline{S}_h}{\overline{\Delta P}_c} \right)^2 \left(\frac{\widehat{U}_{S_h}}{\overline{S}_h} \right)^2 \end{array} \right]^{1/2} \tag{27}$$

$$\frac{\widehat{U}_{OC}}{\overline{OC}} = \left[\begin{array}{l} \left(\frac{\partial OC}{\partial \dot{m}_c} \frac{\overline{\dot{m}}_c}{\overline{OC}} \right)^2 \left(\frac{\widehat{U}_{\dot{m}_c}}{\overline{\dot{m}}_c} \right)^2 + \left(\frac{\partial OC}{\partial \dot{m}_h} \frac{\overline{\dot{m}}_h}{\overline{OC}} \right)^2 \left(\frac{\widehat{U}_{\dot{m}_h}}{\overline{\dot{m}}_h} \right)^2 + \left(\frac{\partial OC}{\partial S_c} \frac{\overline{S}_c}{\overline{OC}} \right)^2 \left(\frac{\widehat{U}_{S_c}}{\overline{S}_c} \right)^2 \\ + \left(\frac{\partial OC}{\partial \eta_p} \frac{\overline{\eta}_p}{\overline{OC}} \right)^2 \left(\frac{\widehat{U}_{\eta_p}}{\overline{\eta}_p} \right)^2 + \left(\frac{\partial OC}{\partial i} \frac{\overline{i}}{\overline{OC}} \right)^2 \left(\frac{\widehat{U}_{i}}{\overline{i}} \right)^2 + \left(\frac{\partial OC}{\partial C_{ele}} \frac{\overline{C}_{ele}}{\overline{OC}} \right)^2 \left(\frac{\widehat{U}_{C_{ele}}}{\overline{C}_{ele}} \right)^2 \end{array} \right]^{1/2} \tag{28}$$

$$\frac{\widehat{U}_{C_{c,o}}}{\overline{C}_{c,o}} = \left[\begin{array}{l} \left(\frac{\partial C_{c,o}}{\partial \dot{m}_c} \frac{\overline{\dot{m}}_c}{\overline{C}_{c,o}} \right)^2 \left(\frac{\widehat{U}_{\dot{m}_c}}{\overline{\dot{m}}_c} \right)^2 + \left(\frac{\partial C_{c,o}}{\partial \dot{m}_h} \frac{\overline{\dot{m}}_h}{\overline{C}_{c,o}} \right)^2 \left(\frac{\widehat{U}_{\dot{m}_h}}{\overline{\dot{m}}_h} \right)^2 + \left(\frac{\partial C_{c,o}}{\partial T_{h,i}} \frac{\overline{T}_{h,i}}{\overline{C}_{c,o}} \right)^2 \left(\frac{\widehat{U}_{T_{c,i}}}{\overline{T}_{c,i}} \right)^2 \\ + \left(\frac{\partial C_{c,o}}{\partial S_c} \frac{\overline{S}_c}{\overline{C}_{c,o}} \right)^2 \left(\frac{\widehat{U}_{S_c}}{\overline{S}_c} \right)^2 + \left(\frac{\partial C_{c,o}}{\partial \eta_p} \frac{\overline{\eta}_p}{\overline{C}_{c,o}} \right)^2 \left(\frac{\widehat{U}_{\eta_p}}{\overline{\eta}_p} \right)^2 + \left(\frac{\partial C_{c,o}}{\partial i} \frac{\overline{i}}{\overline{C}_{c,o}} \right)^2 \left(\frac{\widehat{U}_{i}}{\overline{i}} \right)^2 \\ + \left(\frac{\partial C_{c,o}}{\partial C_{ele}} \frac{\overline{C}_{ele}}{\overline{C}_{c,o}} \right)^2 \left(\frac{\widehat{U}_{C_{ele}}}{\overline{C}_{ele}} \right)^2 + \left(\frac{\partial C_{c,o}}{\partial C_{index}} \frac{\overline{C}_{index}}{\overline{C}_{c,o}} \right)^2 \left(\frac{\widehat{U}_{C_{index}}}{\overline{C}_{index}} \right)^2 \end{array} \right]^{1/2} \tag{29}$$

Where in the above equations the parameters correspond to the following: $\widehat{U}_{h_c}$: uncertainty in cold side heat transfer coefficient, $\widehat{U}_{\Delta P_c}$: uncertainty in cold side pressure drop, $\overline{\Delta P_c}$: nominal value of the cold side pressure drop, $\widehat{U}_{OC}$: uncertainty in operating cost, $\overline{OC}$: nominal value of the operating cost, $\widehat{U}_{C_{c,o}}$: uncertainty in the cold fluid outlet stream cost, $\overline{C_{c,o}}$: nominal value of the cold fluid outlet stream cost, $\overline{h}_c$: nominal value of cold side heat transfer coefficient, $\widehat{U}_{\dot{m}_c}$: uncertainty in cold side flow rate, $\overline{\dot{m}}_c$: nominal value of cold side flow rate, $\overline{\dot{m}}_h$: nominal value of hot side flow rate, $\widehat{U}_{\dot{m}_h}$: uncertainty in hot side flow rate, $\overline{T}_{c,i}$: nominal value of cold fluid inlet temperature, $\widehat{U}_{T_{c,i}}$: uncertainty in cold side inlet temperature, $\overline{T}_{h,i}$: nominal value of hot fluid inlet temperature, $\widehat{U}_{T_{h,i}}$: uncertainty in hot side inlet temperature, $\overline{S}_c$: nominal value of the cold fluid salinity, $\widehat{U}_{S_c}$: uncertainty in the cold fluid salinity,

$\overline{S}_h$: nominal value of the hot fluid salinity, $\widehat{U}_{S_h}$: perturbation in the hot fluid salinity, $\widehat{U}_{\eta_p}$: uncertainty in the pump efficiency value, $\overline{\eta}_p$: nominal value of the pump efficiency, $\widehat{U}_i$: uncertainty in the interest rate, $\overline{i}$: nominal value of the interest rate, $\widehat{U}_{C_{ele}}$: uncertainty in the the electricity cost, $\overline{C}_{ele}$: nominal value of the electricity cost, $\widehat{U}_{C_{index}}$: uncertainty in the cost index factor, $\overline{C}_{index}$: nominal value of the cost index factor.

The relative contribution (RC) is an important parameter in a normalized sensitivity analysis that is used to identify the variable with dominant uncertainty contribution through combining the sensitivity coefficient (SC) with the actual uncertainty. It can calculate as [35].

$$RC = \frac{\left(\frac{\partial Y}{\partial X_j}\widehat{U}_{Xj}\right)^2}{\widehat{U}_Y^2} \tag{30}$$

The working of normalized sensitivity analysis is quite simple. **Figure 2** represents the working methodology of normalized sensitivity analysis. At the start, all the input variables and output performance variables are selected. After that, the uncertainty/perturbation is selected that is generally 1% of the nominal value. In the next step, the partial derivative is taken for each output variable against the various input parameters. After the partial derivate of each variable, the sensitivity coefficient is calculated by using Eq. (23) for all the output variables. In the next step, the total uncertainty and normalized sensitivity of the output variable are calculated by using Eqs. (24) and (25). In the end, derived all the most significant, critical, and dominant input variables in terms of NSC and RC by using Eqs. (26)–(30).

4. Experimental validation of the numerical model

The normalized sensitivity and exergoeconomic techniques are applied on a preheater (plate heat exchanger) of SEE-MVC based-thermal desalination system for which the input data is already summarized in **Table 1**.

For the analysis purpose, a numerical model is developed on Engineering Equation Solver (EES) based using the governing equation mentioned above for which the solution flow chart is presented in **Figure 3**. After that, the developed numerical code is validated with the laboratory/experimental readings from a small-scale PHX as illustrated in **Figure 4**. The specifications of the laboratory scale PHX are mentioned in our previous study [17]. Then, the experiment is carried out for two different operating conditions. For each scenario, the experimental setup is operated for 35 minutes, and readings are saved through a data acquisition system (edibon SCADA) when the system becomes stable. After that, the experimental data are compared with numerical data, as shown in **Figure 5**. The numerical and experimental readings have very close values, which shows the accuracy of the numerical data.

4.1 Normalized sensitivity analysis in terms of NSC and RC

The analysis is carried to identify the most critical and crucial input variable that affects the selected output performance parameters. The desired output performance parameters are local cold side heat transfer coefficient, cold side pressure drop, operational cost, and product cost of the cold stream. **Figure 6** presents the

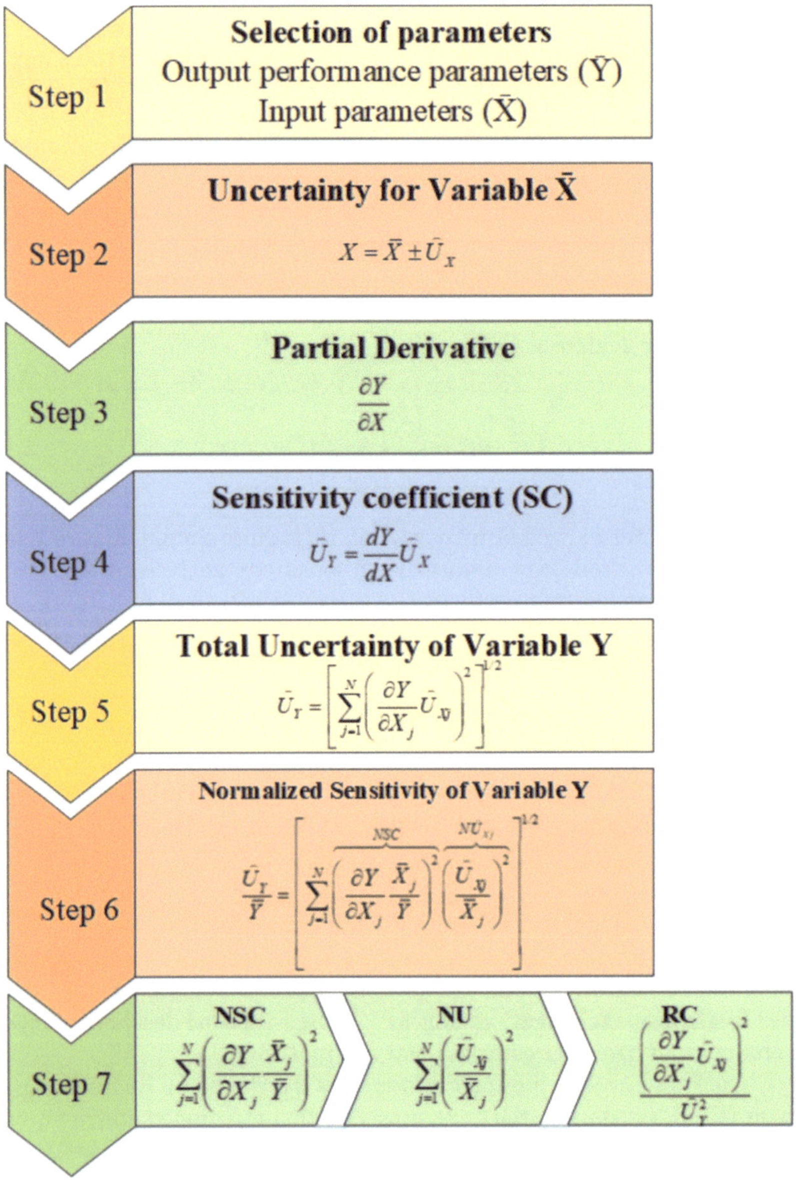

Figure 2.
Working flow chart of normalized sensitivity analysis.

sensitivity analysis results from Normalized Sensitivity Coefficient (NSC) and Relative Contribution (RC). From **Figure 6a**, it can be concluded that for the local heat transfer coefficient, the most crucial variables in terms of NSC are in the following order: cold side mass flow rate $\dot{m}_c$ > inlet temperature of cold side $T_{c,i}$ > salinity of cold side S_c while the RC is highest for cold side mas flow rate $\dot{m}_c$ with ~88% dominancy followed by inlet temperature of cold side with ~11.7% and salinity with ~0.05%. Likewise, for the cold side pressure drop ΔP_c, the most significant variable is $\dot{m}_c$ followed by $T_{c,i}$ while their corresponding RC is 99.6% and 0.4%, respectively as shown in **Figure 6b**. Similarly, from the monetary point of view, the operation cost (OC) highlights that the most influential input variables are $\dot{m}_c$ followed by $\dot{m}_h$, C_{ele}, i, and η_p. The RC is dominated by C_{ele}, with ~86.2% followed by i with ~8.94%,

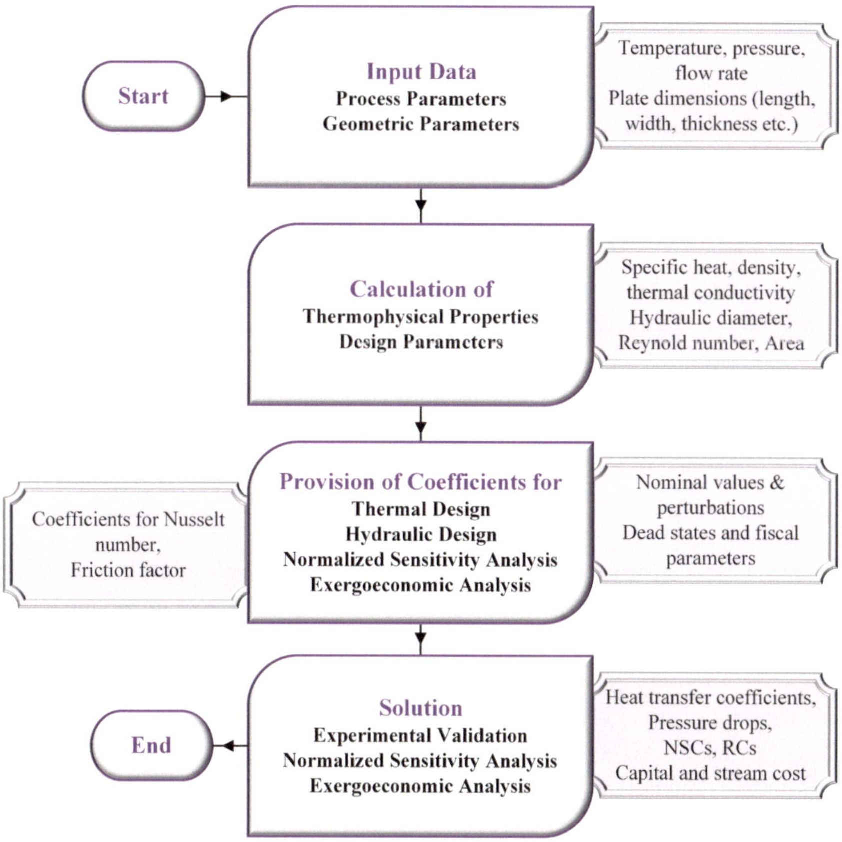

Figure 3.
Solution flow chart for numerical analysis.

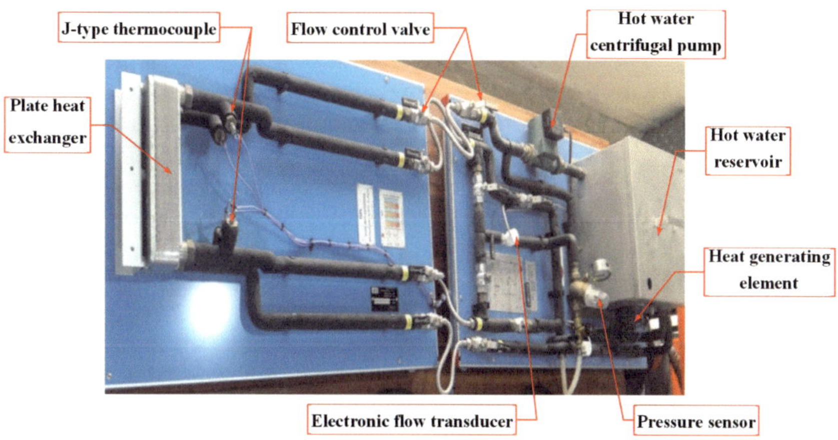

Figure 4.
Experimental test setup.

$\dot{m}_c$ with ~1.88%, $\dot{m}_h$ with ~1.84%, and η_p with ~ 1.15% as illustrated in **Figure 6c**. **Figure 6d** highlights the results of the product cost of the cold stream $C_{c,o}$. The most critical variables in terms of NSC are cost index C_{index} followed by $i, T_{h,i}, \eta_p, \dot{m}_c, \dot{m}_h$ and C_{ele} while the RC is maximum for the inflation rate i with ~95.5%.

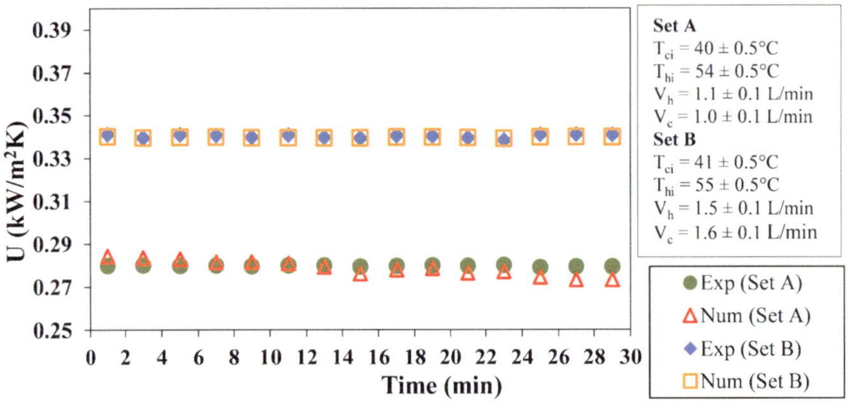

Figure 5.
Model validation with experimental data [17].

Overall, it was observed that the exergoeconomic analysis of PHX is affected by both fascial and process variables. Therefore, fascial parameters must consider equally while designing/analyzing PHX.

4.2 Exergoeconomic analysis

The thermal–hydraulic performance of PHXs is significantly affected by plate chevron angle (β) and mass flow rate [17]. The heat transfer coefficient and pressure drop of the cold stream are increased by varying the Reynold number (Re). However, the rise in heat transfer coefficient is desirable, but the rise in pressure drop is not favorable from a monetary viewpoint. Therefore, the comprehensive parameters (h/ΔP) are calculated to provide a reasonable estimate of heat transfer per unit pressure drop.

From **Figure 7**, the comprehensive performance parameters are declined with the increasing Reynold number. This is because with increasing Reynold number, the pressure drop increased at a higher-order rise compared to the heat transfer coefficient. Furthermore, the analysis is carried out for different chevron angles (β). It can be observed, the h/ΔP is highest for β = 60° followed by β = 50° > 45° > 30°. This is because the pressure drop faces less resistance at a high chevron angle. Meanwhile, from the economic viewpoint, the operation cost (OC) increased as the Reynold number increased. This is because, at the high6Reynold number, the pressure drop is increased which increased the energy consumption and ultimately the pumping power. The operational cost is highest for the chevron angle β = 30° and lowest for chevron angle β = 60° due to low-pressure loss.

Similarly, the product cost of the cold stream $C_{c,o}$ is also increased by varying Reynold number due to increased unit cost of electricity because, at a high Reynold number (flow rate), the pumping power is increased with consumes more energy compared to low-pressure drop. Furthermore, the outlet cost is highest for chevron angle β = 30° followed by β = 45° > 50° > 60°. This is because at a high chevron angle the pressure losses are low as illustrated **Figure 8**.

The traditional analysis is majorly focused on evaluating the consequence of both process and geometric variables. However, in recent studies, the combined analysis of fiscal and process variables gained remarkable importance on the exergoeconomic performance [7, 24]. The primary reason is that the system operating with different economic variables i.e., interest rate, electricity cost, and intake

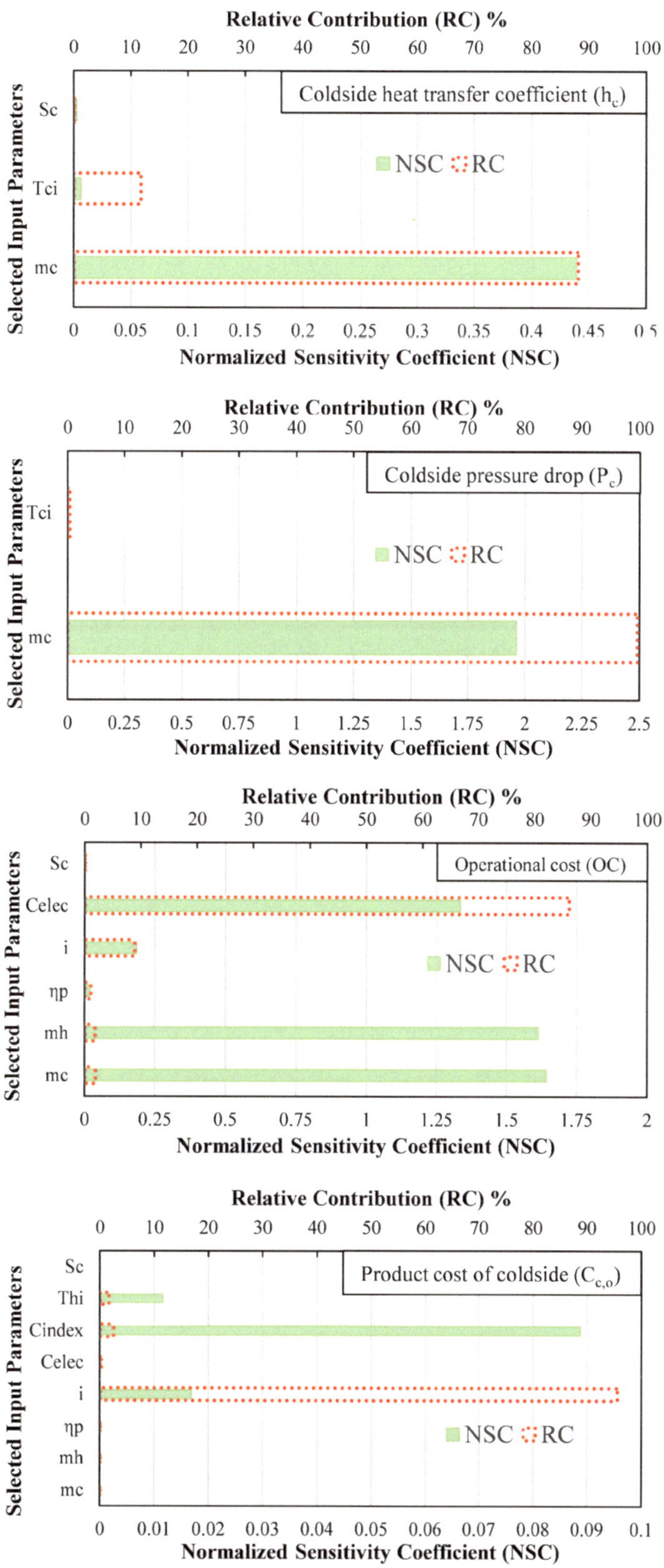

Figure 6.
Normalized Sensitivity analysis results for performance parameters of (a) heat transfer coefficient (h_c) (b) pressure drop (ΔP_c) (c) operational cost (OC) (d) product Cost ($C_{c,o}$).

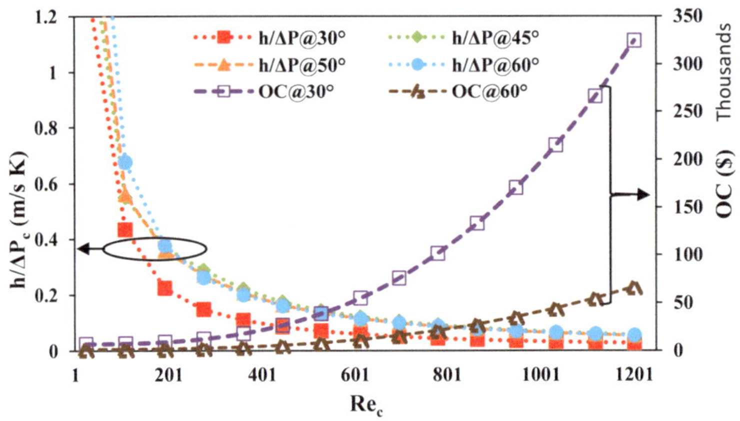

Figure 7.
Effect of Reynold number on heat transfer per unit pressure drop ($h/\Delta P$) and operational cost (OC).

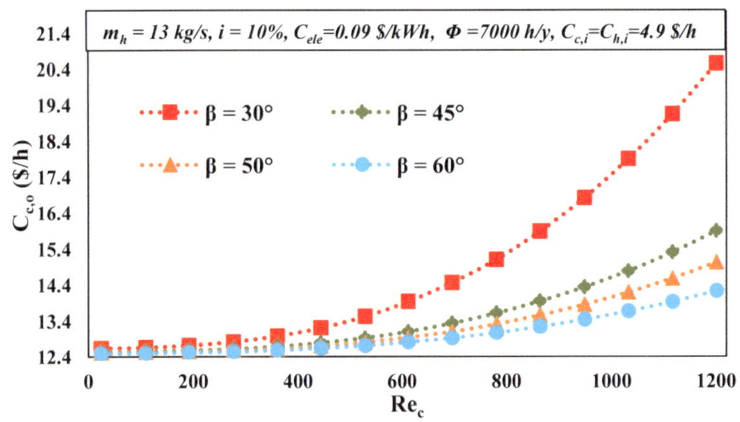

Figure 8.
Effect of Reynold number rate on the outlet cold stream cost ($\dot{C}_{c,o}$).

chemical cost would have different operation cost (OC) with like thermal and hydraulic performance [6, 18].

Therefore, an economic analysis is conducted for various economic policies over time as the importance of fiscal parameters is observed on performance parameters by sensitivity analysis as well in the above section. The cold stream product cost $C_{c,o}$ increased by varying the interest rate and electricity cost, as illustrated in **Figure 9a** and **b**. For example, by varying the inflate rate and from 1 to 14%, the $C_{c,o}$ increased ~17.7% for chevron angle β = 30°. Likewise, for the same chevron angle, the product cost $C_{c,o}$ increased ~3.80% by varying the electricity cost from 0.01 to 0.15 $/kWh. Furthermore, the outlet cost of the cold stream is highest for the β = 30° and lowest for the β = 60° for both interest rate and electricity cost.

An exergoeconomic flow diagram is a noteworthy pictorial demonstration of the thermo-economics output at every significant position of the system. It presents the economics and exergy of all streams at important points, i.e., inlet and outlets of each section of the large system. The visual representation is very substantial for the system with the multiple components to recognize how efficiently the induvial components are working from an economic and exergetic point of view. For the current case study, **Figure 10** demonstrates the exergoeconomic flow diagram.

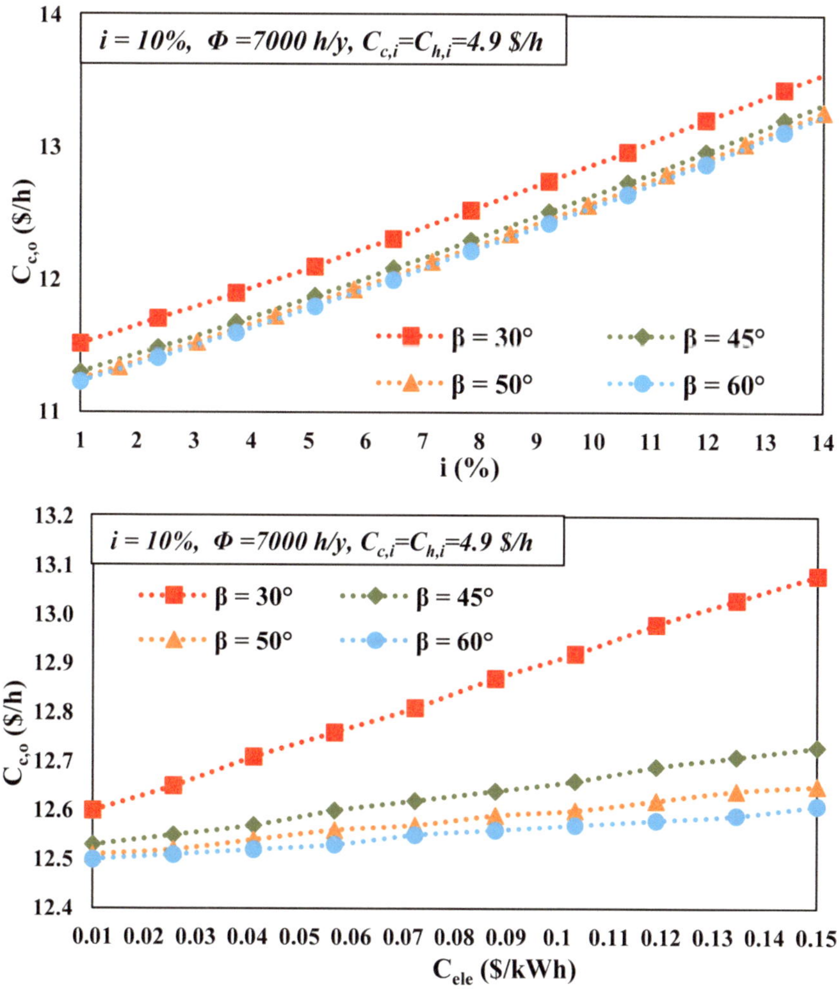

Figure 9.
Effect of monetary (a) cold water product cost ($\dot{C}_{c,o}$) against inflation rate, and (b) cold water product cost ($\dot{C}_{c,o}$) against electricity cost.

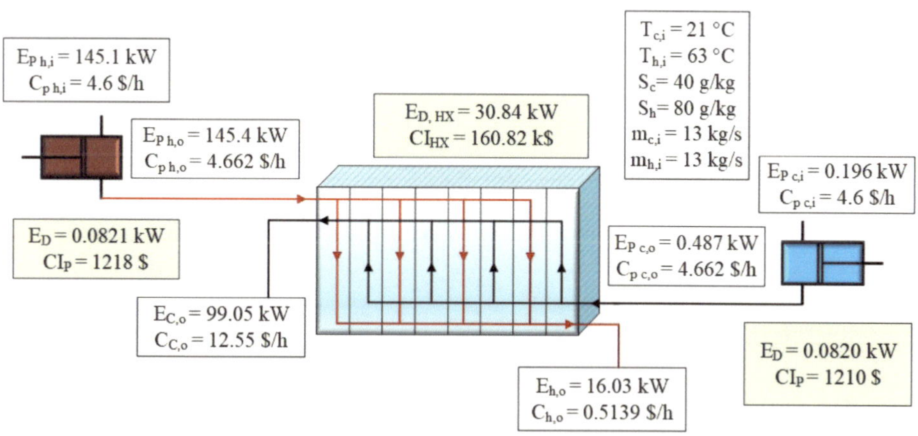

Figure 10.
Exergy-cost flow diagram for current PHX arrangement.

5. Concluding remarks

A corrugated plate heat exchanger (PHX) is examined as a preheater in SEE-MVC based-thermal desalination system to preheat the intake feedwater using the hot waste brine stream. The system is examined from the thermal, hydraulics, and economics point of view. For the case study, the EES-based numerical code is developed using governing equations. After that, the experimental data is used to validate the developed numerical model. Furthermore, sensitivity analysis is conducted in form of NSC and RC to classify the influential input variables. After that, the one-factor-at-a-time (OFAT) technique is used for the detailed parametric analysis to recognize the effect of influential variables. In the end, the exergoeconomic flow diagram is demonstrated to compute the exergies and product cost of the stream at each component of the system. The output of the current case study is as follows.

- The sensitivity analysis highlights that the utmost critical input variables in form of NSC are cold water mass flow rate followed by cold water inlet temperature, and salinity for the local cold water heat transfer coefficient. Similarly, the most critical parameters for the cold side pressure drop are the cold-water mass flow rate followed by the cold-water inlet temperature. Furthermore, the operation cost (OC), the most critical input variable are mass flow of cold water > mass flow of hot water > electricity cost > interest rate > and efficiency of the pump while the cold water outlet cost, the critical variables are cost index > inflation rate > inlet temperature of hot > efficiency of the pump > mass flow rate of water > mass flow rate of hot water >unit cost of electricity.

- The parametric analysis reflects that the comprehensive parameter ($h/\Delta P$) is decreased with an increase of Reynold number due to higher-order increment in pressure drop. Likewise, the operational cost (OC) and cold stream of outlet cost are increased because at high Reynold number, the pressure losses are increased which consume more energy and ultimately increase the pumping power to maintain the desired pressure and overcome the losses. The OC and cold fluid outlet cost is highest for the $\beta = 30°$ and lowermost for $\beta = 60°$ because at a high chevron angle, the pressure loss is low.

- The cold stream outlet cost increased by ~17.7% and ~3.80% by increased the inflation rate and unit cost of electricity respectively for the $\beta = 30°$.

Acknowledgements

The authors would like to thank KAUST Saudi Arabia and Northumbria University UK under reference # RDF20/EE/MCE/SHAHZAD for funding this research.

Conflict of interest

The authors declare no known conflict of interest.

Abbreviations

CI	capital investment
CEPCI	chemical engineering plant cost index
FOC	factor of cleanliness
EF	enlargement factor
HX	heat exchange
IF	installation factor
LMTD	log mean temperature difference
MED	multi-effect desalination
MSF	multistage flash
MVC	mechanical vapor compression
NSC	normalized sensitivity coefficients
PHXs	plate heat exchangers
OSD	over surface design
OFAT	one-factor-at-a-time
OC	running/operational cost
RC	relative contribution
TVC	thermal vapor compression

Nomenclature

A	heat transfer area, m^2
A_e	effective area, m^2
A_p	projected plate area, m^2
A_{sp}	single plate area, m^2
B	mean channel width, m
C_h	constant parameter for calculation of Nusselt number in Eq. (1)
$\dot{C}$	Outlet/product cost, (\$/h)
C_{total}	total equipment cost, \$
C_y	yearly current cost, \$/y
C_{ele}	Electricity cost, \$/kWh
C_{index}	cost index factor
D_p	diameter of port, m
D_{hyd}	hydraulic diameter, m
$\overline{EX}$	specific exergy, kJ/kg
ff	friction factor for pressure drop calculation
ν_{chl}	mass velocity per channel, kg/m^2s
h	heat transfer coefficient (local), W/m2K
h'	enthalpy, kJ
i	inflation/interest rate, %
k	thermal conductivity, W/mK
K_p	constant variable for friction factor calculation in **Table 3**
L_c	compressed plate length, m
L_h	length of horizontal port, m
L_p	vertical port distance from between port ends, m
L_v	vertical port distance between port centers, m
L_w	effective channel width, m
$\dot{m}$	mass flow rate, kg/s
Nu	Nusselt number

n_y	equipment life, year
N_e	effective number of plates
N_p	number of flow passes
N_{tb}	number of HX plates
N_{cpp}	number of flow channels per pass
PP	plate pitch, m
P_{Power}	pumping power, W
ΔP	pressure drop, Pa
Pr	Prandtl number
r	capital recovery factor
Re	Reynolds number
$R_{fo,\ total}$	total fouling resistance, m^2 K/W
S	salinity, g/kg
s	entropy, J/K
T	temperature, °C
t_{plate}	thickness of plate, m
U	global/overall heat transfer coefficient, W/m^2K
V	velocity of fluid, m/s
v_s	specific volume, m^3/kg
$\dot{W}_p$	work of pump, kW
$\dot{E}$	exergy flow rate, kW
$\dot{E}_D$	total exergy destruction, kW
$\dot{\Gamma}$	yearly capital investment rate, $/y

Greek Symbols

ς	rate of fixed cost, $/s
β	chevron angle, deg.
Δ	variation in magnitude
∂	partial
ρ	density, kg/m^3
μ	viscosity, kg/ms
Φ	plant availability/operating hours, hour/year
η_p	Pump efficiency

Subscripts

0	dead state
B	brine
c	cold water
cl	clean
c,i	cold inlet
c,o	cold outlet
chl	per channel
fo	fouled
h	hot
h,i	hot inlet
h,o	hot outlet
i	in
man	manifold

o	out
p	pump
po	port
SW	Seawater
tot	total
w	wall

Superscripts

m	constant parameter for calculation of friction factor in **Table 3**
n	constant parameter for calculation of Nusselt number in Eq. (1)
w	wall

Author details

Muhammad Ahmad Jamil[1*], Talha S. Goraya[2], Haseeb Yaqoob[2], Kim Choon Ng[3], Muhammad Wakil Shahzad[1] and Syed M. Zubair[4]

1 Department of Mechanical and Construction Engineering, Northumbria University, Newcastle Upon Tyne, United Kingdom

2 Department of Mechanical Engineering, Khwaja Fareed University of Engineering and Information Technology, Rahim Yar Khan, Pakistan

3 Water Desalination and Reuse Center, King Abdullah University of Science and Technology, Thuwal, Saudi Arabia

4 Department of Mechanical Engineering, King Fahd University of Petroleum and Minerals, Dhahran, Saudi Arabia

*Address all correspondence to: muhammad2.ahmad@northumbria.ac.uk

References

[1] Abid, A.; Jamil, M. A.; Sabah, N. us; Farooq, M. U.; Yaqoob, H.; Khan, L. A.; Shahzad, M. W. Exergoeconomic Optimization of a Forward Feed Multi-Effect Desalination System with and without Energy Recovery. Desalination, 2020, 499 (July 2020), 114808. DOI: 10.1016/j.desal.2020.114808.

[2] Shahzad MW, Burhan M, Ybyraiymkul D, Ng KC. Desalination processes' efficiency and future roadmap. Entropy. 2019;**21**(1):84. DOI: 10.3390/e21010084

[3] Jamil, M. A.; Shahzad, M. W.; Zubair, S. M. A Comprehensive framework for Thermoeconomic analysis of desalination systems. Energy Convers. Manag., 2020, 222 (June), 113188. DOI:10.1016/j.enconman.2020.113188.

[4] Chitgar N, Emadi MA, Chitsaz A, Rosen MA. Investigation of a novel multigeneration system driven by a SOFC for electricity and fresh water production. Energy Convers. Manag. 2019;**196**(June):296-310. DOI: 10.1016/j.enconman.2019.06.006.

[5] Elsayed ML, Mesalhy O, Mohammed RH, Chow L. C. Transient Performance of MED Processes with Di Ff Erent Feed Con Fi Gurations. 2018; **438**(March):37-53. DOI: 10.1016/j.desal.2018.03.016

[6] Jamil MA, Zubair SM. Effect of feed flow arrangement and number of evaporators on the performance of multi-effect mechanical vapor compression desalination systems. Desalination. September 2017; **2018**(429):76-87. DOI: 10.1016/j.desal.2017.12.007

[7] Jamil MA, Zubair SM. Design and analysis of a forward feed multi-effect mechanical vapor compression desalination system: An Exergo-economic approach. Energy. 2017;**140**: 1107-1120. DOI: 10.1016/j.energy.2017.08.053

[8] Zhou Y, Shi C, Dong G. Analysis of a mechanical vapor recompression wastewater distillation system. Desalination. 2014;**353**:91-97. DOI: 10.1016/j.desal.2014.09.013

[9] Ettouney H. Design of Single-Effect Mechanical Vapor Compression. Desalination. 2006;**190**:1-15. DOI: 10.1016/j.desal.2005.08.003

[10] Elsayed ML, Mesalhy O, Mohammed RH, Chow LC. Transient performance of MED processes with different feed configurations. Desalination. December 2017; **2018**(438):37-53. DOI: 10.1016/j.desal.2018.03.016

[11] Ettouney HM, El-Dessouky HT. Fundamentals of salt water. Desalination. 2002

[12] Abdelkader BA, Jamil MA, Zubair SM. Thermal-hydraulic characteristics of helical baffle Shell-and-tube heat exchangers. Heat Transf. Eng. 2019. DOI: 10.1080/01457632.2019.1611135

[13] Nilpueng K, Keawkamrop T, Ahn HS, Wongwises S. Effect of Chevron angle and surface roughness on thermal performance of single-phase water flow inside a plate heat exchanger. Int. Commun. Heat Mass Transf. 2018;**91**:201-209. DOI: 10.1016/j.icheatmasstransfer.2017.12.009

[14] Shon BH, Jung CW, Kwon OJ, Choi CK, Kang YT. Characteristics on condensation heat transfer and pressure drop for a low GWP refrigerant in brazed plate heat exchanger. Int. J. Heat Mass Transf. 2018;**122**:1272-1282. DOI: 10.1016/j.ijheatmasstransfer.2018.02.077.

[15] Hajabdollahi F, Hajabdollahi Z, Hajabdollahi H. Optimum Design of Gasket Plate Heat Exchanger Using Multimodal Genetic Algorithm. Heat Transf. Res. 2013;**44**(8):761-789. DOI: 10.1615/HeatTransRes.2013006366

[16] Hajabdollahi H, Naderi M, Adimi S. A comparative study on the Shell and tube and gasket-plate heat exchangers: The economic viewpoint. Appl. Therm. Eng. 2016;**92**:271-282. DOI: 10.1016/j.applthermaleng.2015.08.110

[17] Jamil MA, Din ZU, Goraya TS, Yaqoob H, Zubair SM. Thermal-hydraulic characteristics of Gasketed plate heat exchangers as a preheater for thermal desalination systems. Energy Convers. Manag. 2020;**205**(October 2019):112425. DOI: 10.1016/j.enconman.2019.112425

[18] Jamil MA, Zubair SM. On Thermoeconomic analysis of a single-effect mechanical vapor compression desalination system. Desalination. 2017; **420**(July):292-307. DOI: 10.1016/j.desal.2017.07.024.

[19] Kakac S, Liu HHE. Selection, Rating, and Thermal Design. 2nd ed. New York: CRC; 2002

[20] Khairul MA, Alim MA, Mahbubul IM, Saidur R, Hepbasli A, Hossain A. Heat transfer performance and exergy analyses of a corrugated plate heat exchanger using metal oxide Nanofluids. Int. Commun. Heat Mass Transf. 2014;**50**:8-14. DOI: 10.1016/j.icheatmasstransfer.2013.11.006

[21] Mistry KH, McGovern RK, Thiel GP, Summers EK, Zubair SM, Lienhard V, et al. Analysis of desalination technologies. Entropy. 2011;**13**(12):1829-1864. DOI: 10.3390/e13101829.

[22] Nayar KG, Sharqawy MH, Banchik LD, Lienhard V. J. H. Thermophysical properties of seawater: A review and new correlations that include pressure dependence. Desalination. 2016;**390**:1-24. DOI: 10.1016/j.desal.2016.02.024.

[23] Sadeghzadeh H, Ehyaei MA, Rosen MA. Techno-economic optimization of a Shell and tube heat exchanger by genetic and particle swarm algorithms. Energy Convers. Manag. 2015;**93**:84-91. DOI: 10.1016/j.enconman.2015.01.007

[24] Jamil MA, Elmutasim SM, Zubair SM. Exergo-economic analysis of a hybrid humidification dehumidification reverse osmosis (HDH-RO) system operating under different retrofits. Energy Convers. Manag. September 2017;**2018**(158): 286-297. DOI: 10.1016/j.enconman.2017.11.025

[25] Caputo AC, Pelagagge PM, Salini P. Heat exchanger design based on economic optimisation. Appl. Therm. Eng. 2008;**28**(10):1151-1159. DOI: 10.1016/j.applthermaleng.2007.08.010

[26] El-Sayed YM. The Thermoeconomics of Energy Conversions. Amsterdam: Elsevier; 2003

[27] El-Mudir W, El-Bousiffi M, Al-Hengari S. Performance evaluation of a small size TVC desalination plant. Desalination. 2004;**165**:269-279

[28] El-Sayed YM. Designing desalination Systems for Higher Productivity. Desalination. 2001; **134**(1–3):129-158. DOI: 10.1016/S0011-9164(01)00122-9

[29] Jamil MA, Goraya TS, Ng KC, Zubair SM, Bin B, Shahzad MW. Optimizing the energy recovery section in thermal desalination Systems for Improved Thermodynamic, economic, and environmental performance. Int. Commun. Heat Mass Transf. 2021;**124**: 105244. DOI: 10.1016/j.icheatmasstransfer.2021.105244

[30] Zhang C, Liu C, Wang S, Xu X, Li Q. Thermo-economic comparison of subcritical organic Rankine cycle based on different heat exchanger configurations. Energy. 2017;**123**: 728-741. DOI: 10.1016/j.energy.2017.01.132

[31] Li J, Yang Z, Hu S, Yang F, Duan Y. Effects of Shell-and-tube heat exchanger arranged forms on the thermo-economic performance of organic Rankine cycle systems using hydrocarbons. Energy Convers. Manag. 2020;**203**(October 2019):112248. DOI: 10.1016/j.enconman.2019.112248

[32] Vatavuk WM. Updating the CE plant cost index. Chem. Eng. 2002;**1** (January):62-70

[33] Jenkins, S. 2019 Chemical engineering plant cost index annual average https://www.chemengonline.com/2019-chemical-engineering-plant-cost-index-annual-average/ (accessed Apr 9, 2020).

[34] Jamil, M. A.; Xu, B. Bin; Dala, L.; Sultan, M.; Jie, L.; Shahzad, M. W. Experimental and Normalized Sensitivity Based Numerical Analyses of a Novel Humidifier-Assisted Highly Efficient Indirect Evaporative Cooler. Int. Commun. Heat Mass Transf., 2021, 125, 105327. DOI:10.1016/j.icheatmasstransfer.2021.105327.

[35] Jamil, M. A.; Goraya, T. S.; Shahzad, M. W.; Zubair, S. M. Exergoeconomic optimization of a Shell-and-tube heat exchanger. Energy Convers. Manag., 2020, 226 (September), 113462. DOI: 10.1016/j.enconman.2020.113462.

[36] James CA, Taylor RP, Hodge BK. The application of uncertainty analysis to cross-flow heat exchanger performance predictions. Heat Transf. Eng. 1995;**16**(4):50-62. DOI: 10.1080/01457639508939863.

Chapter 2

One-Dimensional Modeling of Triple-Pass Concentric Tube Heat Exchanger in the Parabolic Trough Solar Air Collector

Nguyen Minh Phu and Ngo Thien Tu

Abstract

The parabolic trough solar collector has a very high absorber tube temperature due to the concentration of solar radiation. The high temperature leads to large heat loss to the environment which reduces efficiency of the parabolic trough collector. The heat loss reduction can be obtained by adopting a multi-pass fluid flow arrangement. In this chapter, airflow travels in three passes of the receiver to absorb heat from the glass covers and absorber tube to decrease surface temperatures. 1D mathematical model is developed to evaluate effective efficiency and the temperature distribution of surfaces and fluid. The mathematical modeling is based on air temperature gradients and solved by a numerical integration. Diameter ratios of outer glass to inner glass (r_{23}) and inner glass to absorber tube (r_{12}), Reynolds number (Re), and tube length (L) are varied to examine the efficiency and the temperature distribution. Results showed that the highest efficiency is archived at r_{23} = 1.55 and r_{12} in the range of 1.45 to 1.5. The efficiency increases with Re and decreases with L due to dominant heat transfer in terms of thermohydraulic behavior of a concentrating solar collector. With the optimum ratios, absorber tube temperature can reduce 15 K compared with another case.

Keywords: Concentric tube heat exchanger, 1D modeling, parabolic trough collector, effective efficiency, solar air heater

1. Introduction

Today, energy saving is a matter of great concern due to the depletion of fossil fuels as well as environmental pollution. Faced with that situation, the problem is that we must take advantage of available energy sources and use them for a long time without emitting toxic substances that affect the current living environment. Solar energy is a very abundant, completely free, environmentally friendly and long-lasting source of energy. Besides, it is possible to reduce the amount of toxic waste into the environment and contribute to saving a significant amount of costs when perceiving how to apply solar energy in daily life. Producing hot air by solar energy is an efficient and green solution by using a solar air heater. Flat-plate solar collector is often used in the range of moderately hot air temperature [1–3]. To generate greater air temperature, concentrating solar collectors are used. There are several types of concentrating collector such as linear Fresnel collector, dish collector, and parabolic trough collector. The parabolic trough collector is attributed to the most widespread one [4]. Sketch and pictorial view of a parabolic trough solar collector are shown in **Figures 1** and 2. When the intensity of solar radiation reaches the parabolic surface, solar energy is reflected and concentrated into the focus of the parabola where locates a thermal receiver. Here the heating process takes place to heat up a heat transfer fluid inside the receiver.

Elakhdar et al. [4] used a parabolic trough collector to power the generator of an organic Rankine plant. The results shown that the heat transfer fluid temperature in the receiver tube is up to 430 K and the thermal efficiency of the plant is 0.14. Li and Yuan [5] employed a parabolic trough collector for lighting and heating purposes. The results were reported that the lighting efficiency reaches 16.3%, the thermal efficiency 23.1% and the payback period less than 10 years. Bozorg et al. [6] numerically investigated a parabolic trough collector with nanofluid as a heat transfer media. It is proved that heat transfer, pressure drop, and thermal efficiency increase linearly with Reynolds number. Bellos et al. [7] examined a parabolic

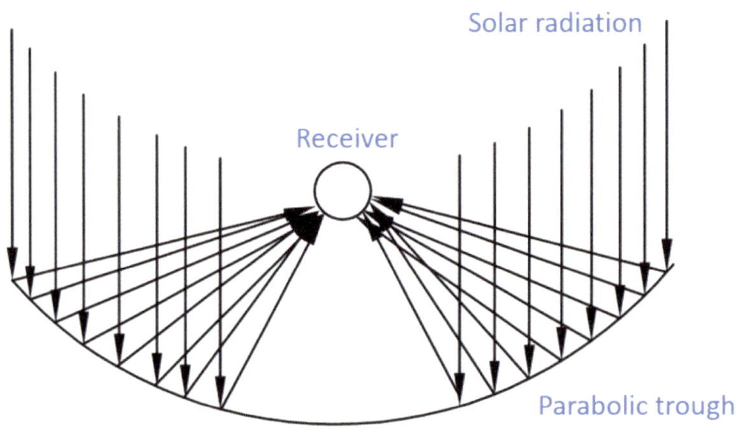

Figure 1.
Parabolic trough solar collector.

Figure 2.
The parabolic trough solar collector being tested at the Ho Chi Minh City University of Technology.

trough collector for thermal energy storage. The thermal efficiency and exergy efficiency of their system indicated 68.7% and 8.5%, respectively. Higher storage tank volume is recommended to yield higher thermal efficiency. Okonkwo et al. [8] studied different configurations of absorber tube including smooth tube, finned tube, twisted tape inserted tube, and converging–diverging tube. The results exhibited that the converging–diverging absorber tube demonstrated the highest exergy efficiency. In addition, the optical losses are the main components of exergy losses. Kaloudis et al. [9] simulated the absorber tube with an inner solid plug as a flow restriction device. In the study, the heat transfer fluid of Al2O3 nanofluid 4% revealed the best collector efficiency. Ray et al. [10] conducted a numerical analysis of selective coatings on the absorber tube. The thermal efficiency can be increased up to 34.6% by the selective coating. Tzivanidis and Bellos [11] used a parabolic trough collector to drive an absorption chiller which is to cool a room. The size of collector and thermal storage tank were determined to meet cooling load of a given space. Ghasemi and Ranjbar [12] compared heat transfer fluids of water and nanofluid inside a absorber tube. They confirmed that the nanofluid can improve heat transfer rate compared with the water. Nain et al. [13] employed the U-tube in a parabolic trough air collector. The tube was covered by the evacuated glass tube to minimize heat loss. The maximum outlet air temperature of 150°C was observed.

From the above literature review, it can be seen that there are many measures to enhance the performance of the parabolic trough solar collector. The present study proposes a triple-pass parabolic trough solar collector configuration so that the heat transfer fluid receives heat from the glass cover and absorber tube surfaces. Thus, the surface temperatures can be reduced leading to increase the collector efficiency. The independent parameters consist of tube diameters, collector length and Reynolds number of the heat transfer fluid to deduce heat transfer characteristics.

2. One-dimensional modeling

Figure 3 shows the schematic diagram of the triple-pass receiver. The receiver consists of an outer glass tube, an inner glass tube and an absorber. The tubes are fitted concentrically to each other so that the airflow moves in the three spaces. Air travels from the outermost annular space followed by the inner annular space and the inside of the absorber tube. **Figure 4** displays a diagram of thermal energy transfer between surfaces and the airflow and a thermal resistance circuit in which the heat conduction resistance of the tubes is neglected. The receiver collects solar radiation from the top half and concentrated radiation from the bottom half. The thermal balance for the glass tube 1 is written as Eq. (1). The solar thermal energy absorbed by the glass is equal to the heat transferred by radiation and convection to

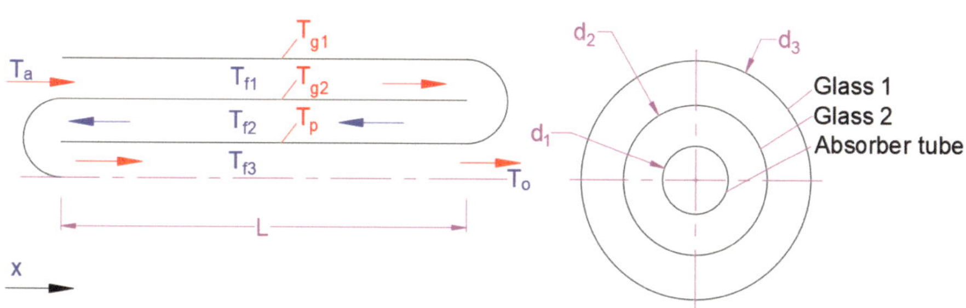

Figure 3.
Details of receiver with triple pass airflow.

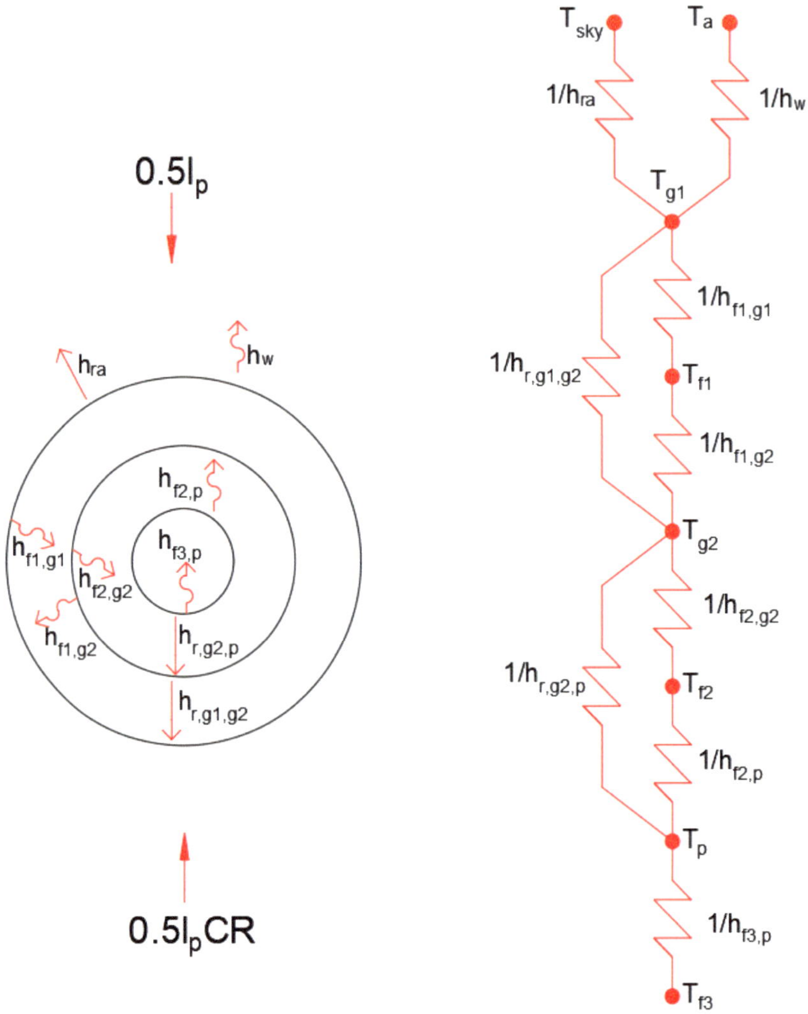

Figure 4.
Energy balance diagram and thermal circuit of triple-pass parabolic trough solar air collector.

the environment, the heat transferred by convection to the fluid in the outer annulus (first pass), and the heat transferred by radiation to the glass tube 2 [6]:

$$I\alpha_{g1}A_{s1} + h_w A_{s3}(T_a - T_{g1}) + h_{ra}A_{s3}(T_{sky} - T_{g1}) + h_{r,g1,g2}A_{s2}(T_{g2} - T_{g1}) + h_{f1,g1}A_{s3}(T_{f1} - T_{g1}) = 0 \quad (1)$$

where I is solar radiation absorbed by the receiver and A_s is surface area of a tube. Solar radiation absorbed by the receiver is the sum of heat flux up and heat flux down [12]:

$$I = (0.5 + 0.5CR)I_p \quad (2)$$

The surface areas can be determined as:

$$A_{s1} = \pi d_1 L \quad (3)$$

$$A_{s2} = \pi d_2 L \quad (4)$$

$$A_{s3} = \pi d_3 L \quad (5)$$

T_{sky} is the sky temperature. $T_{sky} = 0.0552 T_a^{1.5}$ [6].

CR and I_p in Eq. (2) are respectively the concentration ratio and solar radiation per unit area of a surface. CR is defined as [11]:

$$CR = \frac{A_{aperture}}{A_{s1}} \tag{6}$$

where $A_{aperture}$ is aperture area of the parabolic trough.

The temperature variation of the air along the axial direction in the first pass is due to convective heat exchange with the glasses 1 and 2:

$$\frac{dT_{f1}}{dx} = \frac{\pi d_3 h_{f1,g1}(T_{g1} - T_{f1}) + \pi d_2 h_{f1,g2}(T_{g2} - T_{f1})}{\dot{m} c_p} \tag{7}$$

Similar to the glass tube 1, heat exchange of the glass tube 2 is written as:

$$I\tau_{g1}\alpha_{g2}A_{s1} + h_{f1,g2}A_{s2}(T_{f1} - T_{g2}) + h_{f2,g2}A_{s2}(T_{f2} - T_{g2}) + h_{r,g1,g2}A_{s2}(T_{g1} - T_{g2}) \\ + h_{r,g2,p}A_{s1}(T_p - T_{g2}) = 0 \tag{8}$$

The temperature variation of the air in the second pass is given by:

$$\frac{dT_{f2}}{dx} = \frac{\pi d_2 h_{f2,g2}(T_{g2} - T_{f2}) + \pi d_1 h_{f2,p}(T_p - T_{f2})}{\dot{m} c_p} \tag{9}$$

Heat balance of the absorber tube is presented as:

$$I\tau_{g1}\tau_{g2}\alpha_p A_{s1} + h_{f2,p}A_{s1}(T_{f2} - T_p) + h_{r,g2,p}A_{s1}(T_{g2} - T_p) + h_{f3,p}A_{s1}(T_{f3} - T_p) = 0 \tag{10}$$

The temperature variation of the air in the third pass is owing to convection heat transfer with the absorber tube surface:

$$\frac{dT_{f3}}{dx} = \pi d_1 h_{f3,p} \frac{T_p - T_{f3}}{\dot{m} c_p} \tag{11}$$

Convection and radiation heat transfer coefficients in Eqs. (1), (7)–(11) are determined by the following correlations:

- Convection heat transfer coefficient due to wind [6]:

$$h_w = 4 V_w^{0.58} d_3^{-0.42} \tag{12}$$

- Forced convection heat transfer coefficient of the airflow in three passes by Dittus-Boelter correlation [14]:

$$\begin{aligned} h_{f1,g1} &= 0.023 Re_1^{0.8} Pr^{0.4} k/D_{e1} \\ h_{f1,g2} &= h_{f1,g1} \\ h_{f2,g2} &= 0.023 Re_2^{0.8} Pr^{0.4} k/D_{e2} \\ h_{f2,p} &= h_{f2,g2} \\ h_{f3,p} &= 0.023 Re^{0.8} Pr^{0.4} k/d_1 \end{aligned} \tag{13}$$

where Re_1, Re_2 and Re are respectively Reynolds number in passes 1, 2, and 3. Pr is Prandtl number.

$$Re_1 = \rho D_{e1} V_1/\mu \tag{14}$$

$$Re_2 = \rho D_{e2} V_2/\mu \tag{15}$$

$$Re = \rho d_1 V_3/\mu \tag{16}$$

In the above equations, D_e is the equivalent diameter of an annulus. The diameter can be estimated by:

$$D_{e1} = d_3 - d_2 \tag{17}$$

$$D_{e2} = d_2 - d_1 \tag{18}$$

- Radiation heat transfer coefficient of the outer glass to the sky [15]:

$$h_{ra} = \sigma \varepsilon_{g1} \left(T_{g1}^2 + T_{sky}^2\right)\left(T_{g1} + T_{sky}\right) \tag{19}$$

- Radiation heat transfer coefficient of two glass tubes [4]:

$$h_{rg1g2} = \sigma\left(T_{g1}^2 + T_{g2}^2\right)\frac{T_{g1} + T_{g2}}{1/\varepsilon_{g2} + \left(1 - \varepsilon_{g1}\right)\frac{A_{s2}/A_{s3}}{\varepsilon_{g1}}} \tag{20}$$

- Radiation heat transfer coefficient of the inner glass tube and the absorber tube [4]:

$$h_{rg2,p} = \sigma\left(T_{g2}^2 + T_p^2\right)\frac{T_{g2} + T_p}{1/\varepsilon_p + \left(1 - \varepsilon_{g2}\right)\frac{A_{s1}/A_{s2}}{\varepsilon_{g2}}} \tag{21}$$

- Air velocities in passes (V_1, V_2, and V_3) can be evaluated by mass conservation as:

$$\dot{m} = \left(\pi\frac{d_3^2}{4} - \pi\frac{d_2^2}{4}\right)\rho V_1 = \left(\pi\frac{d_2^2}{4} - \pi\frac{d_1^2}{4}\right)\rho V_2 = \pi\frac{d_1^2}{4}\rho V_3 \tag{22}$$

The energy amount of air received when passing through the receiver is determined as follows:

$$Q = \dot{m} c_p (T_o - T_a) \tag{23}$$

As the flow rate and receiver size change, the pumping power will also change to transport the fluid. The useful hydraulic energy of the air received from a blower is calculated as follows:

$$P_{flow} = \dot{m}\frac{\Delta P}{\rho} \tag{24}$$

where ΔP is the pressure difference of the air across the receiver. Because of significant length of the receiver, minor losses are omitted. The pressure drop due to friction in three 3 passes can be computed as:

$$\Delta P = \rho f_1 V_1^2 L/D_{e1} + \rho f_2 V_2^2 L/D_{e2} + \rho f_3 V_3^2 L/d_1 \tag{25}$$

where f is the friction factor of the air with the tube surface. The factor is calculated from the Blasius equation as follows [15]:

$$\begin{aligned} f_1 &= 0.059\, Re_1^{-0.2} \\ f_2 &= 0.059\, Re_2^{-0.2} \\ f_3 &= 0.059\, Re^{-0.2} \end{aligned} \tag{26}$$

The effective efficiency of a parabolic trough collector taking into account the heat received, the power dissipated relative to the radiation absorbed by the receiver is calculated as follows:

$$\eta_{eff} = \frac{Q - P_{flow}/C_o}{A_{s1} I} \tag{27}$$

where C_o is thermal energy conversion factor, C_o = 0.2 [16].

In this study, the absorber tube diameter was fixed. The diameter of the glass tubes, the Reynolds number of the air in the absorber tube, and the receiver length varied to investigate the axial temperature distribution and the collector efficiency. The diameters of the glasses change according to the diameter ratios which are defined as follows:

$$r_{23} = d_3/d_2 \tag{28}$$

$$r_{12} = d_2/d_1 \tag{29}$$

Table 1 presents the parameters entered into the mathematical model. The thermophysical parameters of the air (specific heat c_p, thermal conductivity k, density ρ, and dynamic viscosity μ) are estimated at the ambient temperature. Temperature variations of three surfaces and air in three passes can be solved by numerical integrals of temperature gradient equations, i.e., Eqs. (7), (9), and (11). The one-dimensional (1D) computational domain along the tube is divided into control volumes corresponding to a spatial step Δx. Solving the ordinary differential equations can be utilized ode45 function in MATLAB software [17] or integral function in EES software [18]. **Figure 5** presents a comparison of the air flow

Parameter	Value
Concentration ratio	CR = 10
Solar radiation	I_p = 848 W/m^2
Ambient temperature	T_a = 27°C
Absorptivity of glass covers	$\alpha_{g1} = \alpha_{g2} = 0.05$
Absorptivity of absorber tube	$\alpha_p = 0.92$
Emissivity of glass covers	$\varepsilon_{g1} = \varepsilon_{g2} = 0.92$
Emissivity of absorber tube	$\varepsilon_p = 0.92$
Transmissivity of glass cover	$\tau_{g1} = \tau_{g2} = 0.84$
Wind velocity	V_w = 1 m/s
Absorber tube diameter	d_1 = 42 mm

Table 1.
Input parameters.

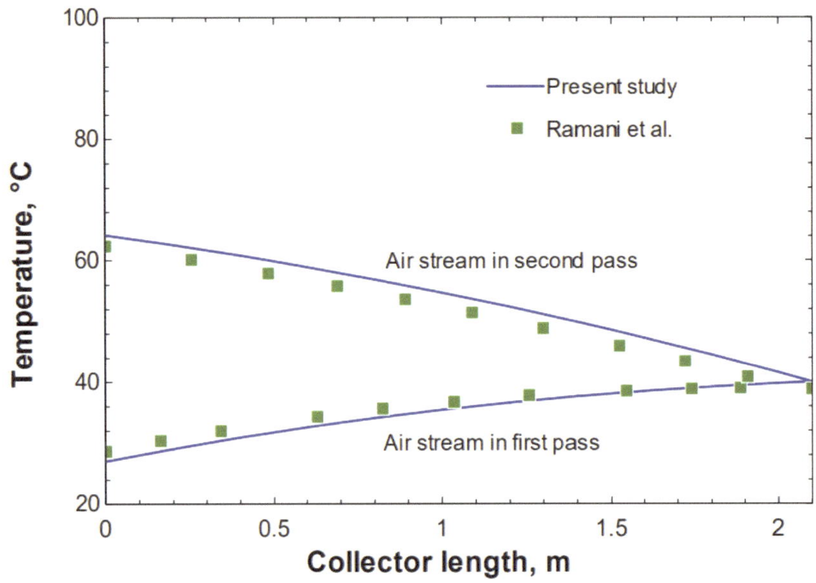

Figure 5.
Confirmation of air temperature along the collector length with published data [19].

temperature variation in a double-pass solar air heater between the results from the code developed in this study and those of the literature [19]. It can be seen that the two results are close to each other and the difference is not significant. Therefore, the model formulation in this study is accuracy to perform the parametric study for the triple pass parabolic trough solar air collector.

3. Results and discussion

This section investigates the influence of Reynolds number in the range 10000 to 16000, tube diameter ratios (r_{23} and r_{12}) from 1.2 to 2, and tube length from 1.5 to 3.5 m on the collector efficiency and temperature distribution of the receiver. **Figure 6** shows the effect of Re and r_{23} on the performance. It can be seen that when the Re number increases, the performance increases. For flat-plate air collectors, efficiency can peak at some Re. However, for a concentrating collector, the thermal duty is very large compared to pumping power. Therefore, when Re increases, the heat transfer rate increases more than increase in the pressure loss penalty. However, the efficiency increases slightly with the Re number because the pumping power increases significantly with the flow rate. At fixed Re, the performance peaks at certain r_{23}. As the ratio increases, the diameter of the glass covers increases, which reduces the heat transfer rate due to the decrease in air velocity. Conversely, increasing this ratio reduces the pressure loss in the two annuli. This trade-off leads to the optimal r_{23}. **Figure 6** shows that the optimal r_{23} value is almost unchanged with Re, i.e., $r_{23,\ opt} = 1.55$. **Figure** 7 presents the effect of the ratio of inner glass tube diameter to absorber tube diameter on the efficiency. The effect of r_{12} can be seen to be more pronounced than that of r_{23}. This is attributed to mainly heat exchange in the second pass due to the high temperatures of the absorber tube and glass 2. The optimum r_{12} varies over a wider range from 1.45 to 1.5. Also, at certain Re number, the performance changes drastically with r_{12}, especially at large Re numbers. At Re = 16000, the efficiency varies from 0.59 to 0.6 with r_{23}, from 0.575 to 0.605 with r_{12}.

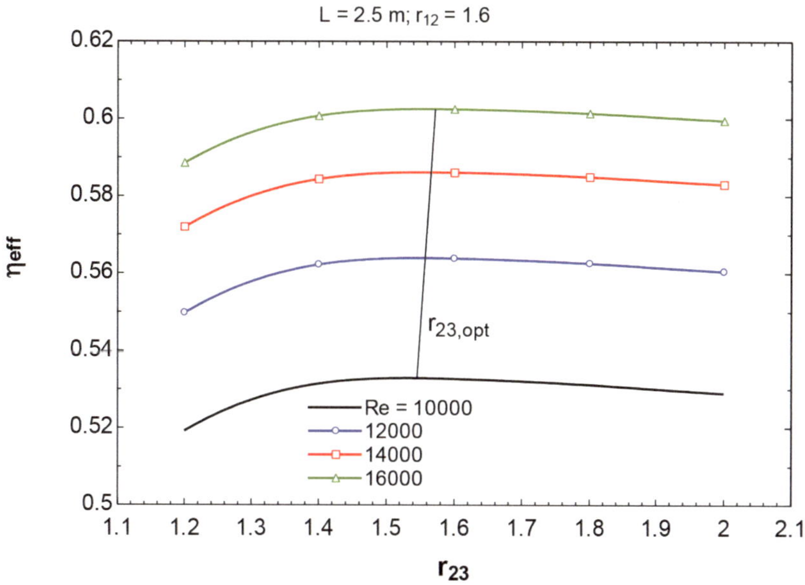

Figure 6.
Effect of Reynolds number and diameter ratio of glasses.

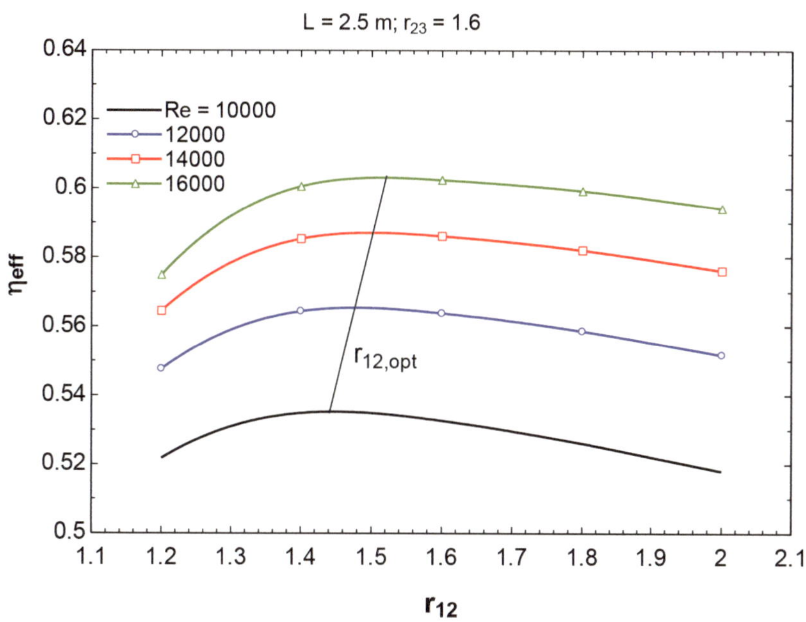

Figure 7.
Effect of Reynolds number and diameter ratio of glass 2 and absorber tube.

Figure 8 depicts the effect of collector length (L) on collector efficiency. It is observed that as the length increases, the efficiency decreases. It is noted that for a concentrating collector, the thermal power is very large compared to the pumping power. Therefore, the optimum L and Re are not found to maximize the efficiency as can be observed in flat-plate solar collectors. As the length increases, at a certain pass, the air temperature increases, reducing the temperature difference between the heat exchanger surfaces and the air temperature. This small temperature difference causes poor heat transfer rate. The reduction in efficiency with the length in the concentrator is also similar to that of the flat plate collector [15].

Figures 9 and **10** illustrate the distributions of the six temperatures along the tube length at Re = 10000. **Figure 9** plots the temperatures at the optimal diameter ratios for this Re, i.e., r_{12} = 1.44 and r_{23} = 1.56. While **Figure 10** represents worst-case temperatures (r_{12} = 2 and r_{23} = 1.2) for the sake of comparison. It is clear that the average temperature of the absorber tube decreases markedly from 160°C (the worst case) to 145°C (the optimum case). This reveals the optimal case owning better heat exchange of the fluid in passes 2 and 3 with absorber tube surface and

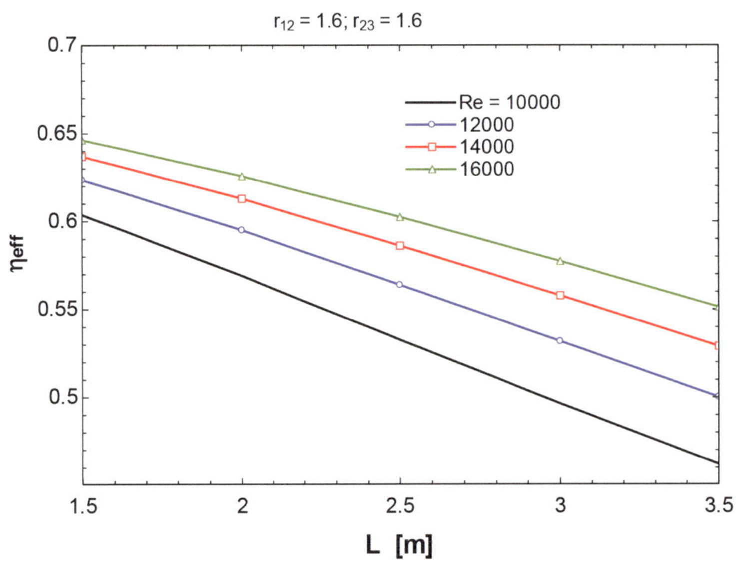

Figure 8.
Effect of Reynolds number and collector length.

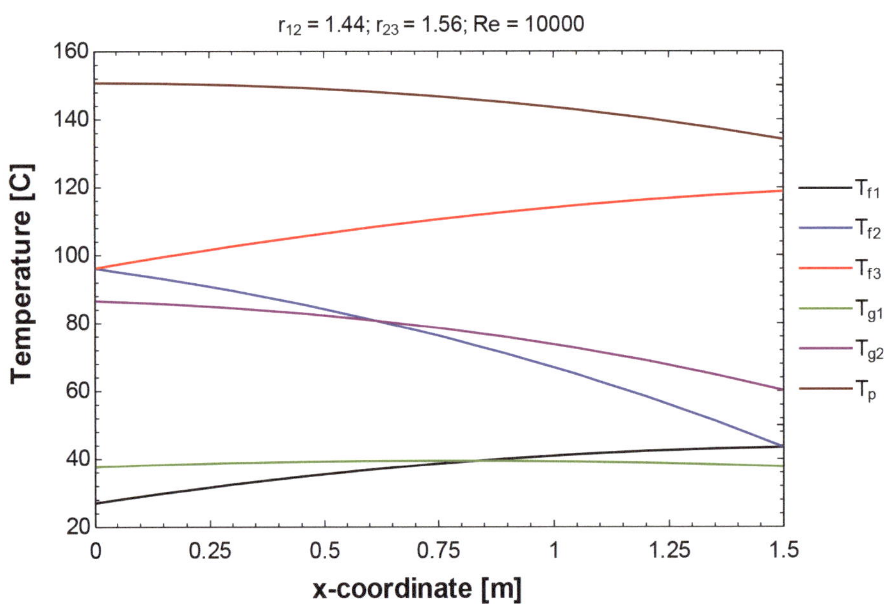

Figure 9.
Local temperature distribution of the airflow in passes and surfaces at optimum ratios.

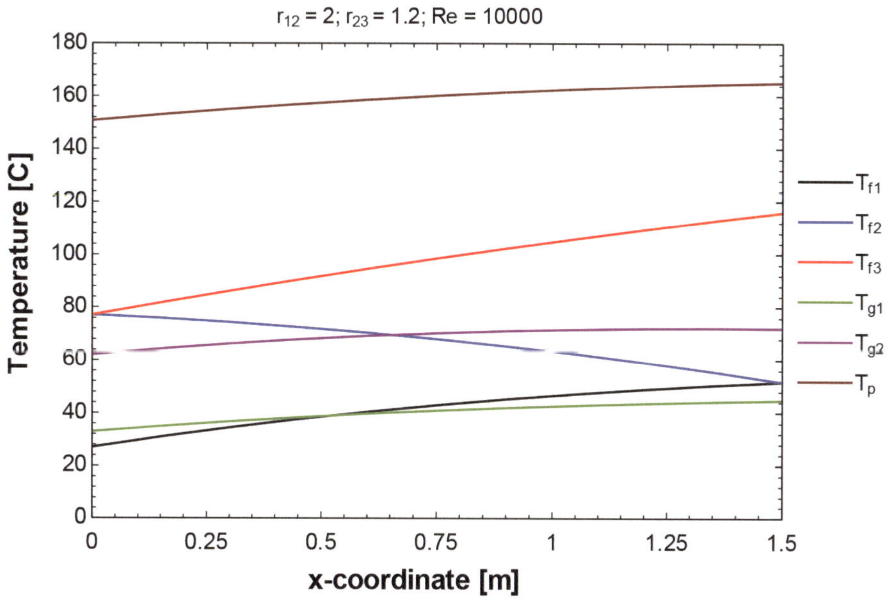

Figure 10.
Local temperature distribution of the airflow in passes and surfaces at the worst ratios.

smaller heat loss. For the optimal case, the absorber tube temperature increases with the air temperature in the second pass. In contrast, the absorber tube temperature increases with the air temperature in third pass for the worst case. Furthermore, the temperature cross between the air in pass 1 and glass 1 occurs in the very short segment from the inlet in the worst case. The temperature differences in pass 2 of **Figures 9** and **10** are 55 K and 28 K, respectively. From **Figures 9** and **10** it can be recognized that there are two temperature crosses between T_{f1} and T_{g1}, and between T_{f2} and T_{g2}. If shorter tubes are used, a temperature cross may not occur. Therefore, it can be concluded that collector efficiency decreases with increasing tube length by means of the current axial temperature analysis.

4. Conclusions

A 1D analytical model has been presented in this chapter to predict the local temperature of fluid and heat exchanger surfaces of the triple-pass parabolic trough solar collector. Some important conclusions are drawn as follows:

- The diameter ratios of outer to inner glasses of 1.55 and the inner glass to absorber tube in the range of 1.45 to 1.5 achieve the greatest effective efficiency.

- The efficiency increases with Reynolds number and decreases with tube length. In other words, optimum Reynolds number and tube length were not found due to the fact that thermal power prevails over pumping power for a concentrating solar collector.

- The absorber tube temperature is reduced up to 15 K with optimal diameter ratios.

- The effect of r_{12} on efficiency is more significant than that of r_{23} due to the strong heat exchange of the airflow in the annular space between the absorber tube and the inner glass tube.

Acknowledgements

This research is funded by the Vietnam National University-Ho Chi Minh City (VNU-HCM) under grant number B2021-20-06.

Conflict of interest

The authors declare no conflict of interest.

Author details

Nguyen Minh Phu[1*] and Ngo Thien Tu[2,3]

1 Faculty of Heat and Refrigeration Engineering, Industrial University of Ho Chi Minh City (IUH), Ho Chi Minh City, Vietnam

2 Faculty of Mechanical Engineering, Ho Chi Minh City University of Technology (HCMUT), Ho Chi Minh City, Vietnam

3 Viet Nam National University, Ho Chi Minh City, Vietnam

*Address all correspondence to: nguyenminhphu@iuh.edu.vn

References

[1] Phu NM, Luan NT. A Review of Energy and Exergy Analyses of a Roughened Solar Air Heater. Journal of Advanced Research in Fluid Mechanics and Thermal Sciences. 2020;77(2): 160-175.

[2] Phu NM, Hap NV. Performance Evaluation of a Solar Air Heater Roughened with Conic-Curve Profile Ribs Based on Efficiencies and Entropy Generation. Arabian Journal for Science and Engineering. 2020;45(11):9023–9035.

[3] Phu NM, Hung HN, Tu NT, Van Hap N. Analytical predictions of exergoeconomic performance of a solar air heater with surface roughness of metal waste. Journal of Thermal Analysis & Calorimetry. 2021;144(5).

[4] Elakhdar M, Landoulsi H, Tashtoush B, Nehdi E, Kairouani L. A combined thermal system of ejector refrigeration and Organic Rankine cycles for power generation using a solar parabolic trough. Energy Conversion and Management. 2019;199:111947.

[5] Li T, Yuan C. An optimal design analysis of a novel parabolic trough lighting and thermal system. International Journal Of Energy Research. 2016;40(9):1193-1206.

[6] Bozorg MV, Doranehgard MH, Hong K, Xiong Q. CFD study of heat transfer and fluid flow in a parabolic trough solar receiver with internal annular porous structure and synthetic oil–Al2O3 nanofluid. Renewable Energy. 2020;145:2598-2614.

[7] Bellos E, Tzivanidis C, Belessiotis V. Daily performance of parabolic trough solar collectors. Solar Energy. 2017;158: 663-678.

[8] Okonkwo EC, Abid M, Ratlamwala TA. Effects of synthetic oil nanofluids and absorber geometries on the exergetic performance of the parabolic trough collector. International Journal of Energy Research. 2018;42 (11):3559-3574.

[9] Kaloudis E, Papanicolaou E, Belessiotis V. Numerical simulations of a parabolic trough solar collector with nanofluid using a two-phase model. Renewable Energy. 2016;97:218-229.

[10] Ray S, Tripathy AK, Sahoo SS, Bindra H. Performance analysis of receiver of parabolic trough solar collector: E ffect of selective coating, vacuum and semitransparent glass cover. International Journal of Energy Research. 2018;42(13):4235-4249.

[11] Tzivanidis C, Bellos E. The use of parabolic trough collectors for solar cooling–A case study for Athens climate. Case Studies in Thermal Engineering. 2016;8:403-413.

[12] Ghasemi SE, Ranjbar A. Effect of nanoparticles in working fluid on thermal performance of solar parabolic trough collector. Journal of Molecular Liquids. 2016;222:156-166.

[13] Nain S, Parinam A, Kajal S. Experimental study and analysis of air heating system using a parabolic trough solar collector. International Journal of Ambient Energy. 2018;39(2):143-146.

[14] Thao PB, Truyen DC, Phu NM. CFD Analysis and Taguchi-Based Optimization of the Thermohydraulic Performance of a Solar Air Heater Duct Baffled on a Back Plate. Applied Sciences. 2021;11(10):4645.

[15] Nguyen Thanh L, Nguyen Minh P. First and Second Law Evaluation of Multipass Flat-Plate Solar Air Collector and Optimization Using Preference Selection Index Method. Mathematical Problems in Engineering. 2021;2021: 5563882.

[16] Matheswaran M, Arjunan T, Somasundaram D. Energetic, exergetic and enviro-economic analysis of parallel pass jet plate solar air heater with artificial roughness. Journal of Thermal Analysis and Calorimetry. 2019;136(1): 5-19.

[17] Phu NM, Lee GS. Characteristics of pressure and force considering friction in a closed cylinder with a holed piston. Journal of Mechanical Science and Technology. 2014;28(6):2409-2415.

[18] Phu NM. A compact EES program to predict the axial temperature distribution in triple-fluid heat exchanger. Science & Technology Development Journal. 2020;3(3): 452-460.

[19] Ramani B, Gupta A, Kumar R. Performance of a double pass solar air collector. Solar energy. 2010;84(11): 1929-1937.

Section 2

Novel Working Fluids

Chapter 3

Magneto-Hydrodynamic Natural Convection Flow in a Concentric Annulus with Ramped Temperature and Ramped Motion of the Boundaries

Khadijah Lawal and Haruna Jibril

Abstract

An unsteady MHD flow of a temperature dependent heat source/sink in an annulus due to ramped motion and ramped temperature of the boundaries has been analyzed. The partial differential equations of the fluid flow are formulated taking into account the ramped temperature and ramped velocity of the inner cylinder. The closed form solution are obtained for three cases of the magnetic field being fixed relative the fluid, cylinder and when the velocity of the magnetic field is less than the velocity of the moving cylinder. The problem is solved using Laplace transform technique to obtain the Laplace domain solution and Riemann sum approximation to obtain the time domain solution. The effect of the governing parameters on the fluid flow are illustrated graphically. It is found that, Hartmann number has a retarding effect on the skin friction at the outer surface of the inner cylinder and mass flow rate. It also decreases fluid velocity for cases $(K = 0.0 \ and \ K = 0.5)$ the reverse effect is noticed for case $(K = 1.0)$. Increase in Hartmann number lead to an increase in skin friction at the inner surface of the outer cylinder for case $(K = 0.0)$ but decreases it for cases $(K = 0.0 \ and \ K = 0.5)$.

Keywords: ramped temperature, ramped motion, magneto-hydrodynamic, natural convection, annulus, heat source/sink

1. Introduction and definition of terms

Magneto-hydrodynamics (MHD) is the study of the motion of electrically conducting fluid. The study of magneto-hydrodynamic plays an important role in agriculture, engineering and petroleum industries. For instance, it may be used to deal with problems such as cooling of nuclear reactor by liquid sodium. The importance of MHD cannot be over emphasized. MHD has applications in many areas like the earth, sun, industry, fusion etc.

To appreciate the importance of fluid dynamics in life demands little more than just a glance around us. In general, life as we know would not exist if there are no fluids and the behavior they exhibit. The water and air we respectively drink and breathe are fluids. In addition, our body fluids are mostly water based. As a matter of fact, our body system is made up of about 75% of fluid which helps in regulating the activities of the body system ranging from body temperature control to waste removal. In a more practical setting, like in our transportation systems, recreation, entertainment (sound from radio speakers) and our sleep (water beds), fluids greatly influence our comfort. It is clear to see from this that engineers need a clear knowledge of fluid behavior to handle many systems of their encounter.

Over the past decades, studies have been carried out on magneto-hydrodynamic natural convection in an annulus under different physical situations and geometry. This is because of its applications in nature, engineering, industries and technologies. These applications include but not limited to underground disposal of radioactive waste materials, storage of foodstuffs, exothermic and/or endothermic chemical reactions, heat removal from nuclear fuel debris, dissociating fluids in packed bed reactors, aerodynamics, geothermal energy extraction, purification of crude oil and spacecraft, MHD generators, MHD flow meters and MHD pump.

- **Magneto-hydrodynamics (MHD):** Magneto-hydrodynamics is the study of electrically conducting fluids in the presence of magnetic and electric fields, example, plasma, liquid metals and salt water.

- **Free or Natural Convection:** Free or Natural convection is the process when a temperature difference produces a density difference which results in mass movement.

- **Forced convection:** Is a mechanism or form of heat transport in which fluid motion is generated by an external source (like suction device fan and pump).

- **Mixed convection:** Is the type of heat transport caused by both natural and force convections.

- **Ramped temperature:** Is the gradual rate of change in temperature over time expressed in degree per second.

- **Ramped motion:** Is a gradual transition in any flow parameter that can be animated from the start value to the end. The length of the ramp behavior in the time line defines the speed of the transition by the behavior's end value.

- **Heat sink:** Is any environment or medium that absorb heat. It decreases the heat of the fluid on the cylinder by an external agent. It is an environment capable of absorbing heat from substance within it and with which it is in thermal contact, without an appreciable change in its own phase.

- **Heat source:** Is any device or natural body that supplies heat. Is the increase of heat of the fluid on the cylinder by an external agent from the place or the environment which heat is obtained.

- **Annulus**: Annulus is the area bounded by two concentric cylinders.

The presence of magnetic field on a Couette flow induces a Lorentz force which either accelerates or decelerates the flow element between the planes which depend on the electrical properties of the plane. The need to control the motion of the boundary layer is one motivation for this study. In many technological phenomena, such as earth core, aeronautics etc. the motion of a system initially start with an accelerated velocity and then after some time moves with almost constant velocity. This prompted us to consider the ramped like motion of a concentric cylinder and analyze the flow formation.

Chandran *et al.* [1] studied natural convection near a vertical plate with ramped wall temperature and they obtained two different solutions, one valid for Prandtl number different from unity and the other for which the Prandtl number is unity. They concluded that the solutions for dimensional velocity and temperature variables depend upon the Prandtl number of the fluid and the expression of the fluid velocity is not uniformly valid for all values of Prandtl number. In their work, heat generating/absorption is absent. However, when the temperature differences are appreciably large, the volumetric heat generation/absorption term may exert strong influence on the heat transfer and as a consequence on the fluid flow as well. Jha *et al.* [2] studied natural convection flow of heat generating or absorbing fluid near a vertical plate with ramped temperature and consider two cases, plate with continuous ramped temperature and the other with isothermal temperature. They concluded that the isothermal case is always higher than the ramped case. The above mentioned works are carried out in the absence of magnetic field. Seth and Ansari [3] considered hydro-magnetic natural convections flow past an impulsively moving vertical plate embedded in a porous medium with ramped wall temperature in the presence of thermal diffusion with heat absorption. Nandkeolyar and Das [4] studied unsteady MHD free convection flow of a heat absorbing dusty fluid past a flat plate with ramped wall temperature. Seth and Nandkeolyar [5] studied MHD natural convection flow with radiative heat transfer past an impulsively moving plate with ramped wall temperature. Again, Seth et al. [6] investigated hydromagnetic natural convection flow with heat and mass transfer of a chemically reacting and heat absorbing fluid past an accelerated moving plate with ramped temperature and ramped surface concentration through a porous medium. Recently, Khadijah and Jibril [7] studied Unsteady MHD natural convection flow of heat generating/absorbing fluid near a vertical plate with ramped temperature and motion. In the same year, Khadijah and Jibril [8] investigated Time dependent MHD natural convection flow of a Heat generating/absorbing fluid near a vertical porous plate with ramped boundary conditions.

Jha and Jibril [9] investigated hydro-magnetic flow due to ramped motion of the boundary. In their work they studied the effect of magnetic field on velocity and skin friction, due to ramped motion of the horizontal plate and it was concluded that the ramped motion are less compared to the constant motion. However, the effect of ramped temperature profile was not considered. Kumar and Singh [10] studied the transient magneto hydrodynamic Couette flow with ramped velocity. The velocity of the magnetic field, applied perpendicular to the plate is taken to be different from the velocity of the lower plate (the lower plate is moving with ramped velocity). It was concluded that the effect of the velocity on the magnetic field is to increase the velocity of the fluid from the upper plate to the lower plate Jha and Jibril [11] studied the effects of transpiration on the MHD flow near a porous plate having ramped motion. In their work they compare flow formation due to ramped motion of porous plate with the flow formation due to constant motion of the porous plate. Jha and Jibril [12] studied the time dependent MHD Couette flow due to ramped motion of one of the boundaries. It was found that velocity and skin friction increases with an increase of Hartman number when the magnetic field is fixed with respect to the moving plate. While the reverse when it is fixed with respect to the fluid. Jha and Jibril [13] investigated the unsteady hydromagnetic Couette flow due to ramped motion of the porous plate. The aforementioned works were carried out on a horizontal plate.

Jha and Apere [14] on the other hand investigated Unsteady MHD Couette flow in an annulus, by applying Riemann-sum approximation approach to obtain the Laplace inversion of their solution in time domain. Jha and Apere [15] studied unsteady MHD two-phase Couette flow of fluid particles suspension in an annulus. In their work, they employed the D'Alermbert method used by Reccebli and kurt in conjunction of Riemann sum approximation method for both cases of the magnetic field being fixed to either the fluid or the moving cylinder to obtain the solution of the problem. Anand [16] investigated the Effect of radial magnetic field on free convective flow over ramped velocity moving vertical inner cylinder with ramped type temperature and concentration. In the same year, Anand [17] studied the effect of radial magnetic field on natural convection flow in alternate conducting vertical concentric annuli with ramped temperature. Anand [18] studied the effect of velocity of applied magnetic field on natural convection over ramped type moving inner cylinder with ramped type temperature solved numerically by using implicit finite difference Crank–Nicolson method. They found out that, when velocity is employed to magnetic field, then effect of magnetic field gets reversed and the effect of velocity of magnetic field get more pronounced with radii ratio. Also, Hartmann number and time parameter have increasing effects on the skin-friction profile. Taiwo [19] investigated the exact solution of MHD natural convection flow in a concentric annulus with heat absorption. It is found that the magnitude of maximum fluid velocity is greater in the case of isothermal heating compared with the constant heat flux case when the gap between the cylinders is less or equal to radius of the inner cylinder. More also, the various values of the non-dimensional heat absorption parameter (A) and the corresponding values of annular gap are almost the same conditions. Other important researchers [20–28] investigated MHD flow under different physical geometry and thermal conditions of the boundaries.

To the best of the authors' knowledge, no studies have been reported concerning the combined effect of constant temperature, heat generating/absorbing parameter, MHD and ramped like motion of the inner cylinder and temperature in a concentric cylinder. The condition involving ramped like cylindrical motion appears in aerodynamics and oil refinement industry. Therefore, it is important to analyze the flow processes and try to understand the function of related mechanics of ramped moving vertical cylinder and ramped temperature.

2. Mathematical formulation

This research considers the time dependent natural convection flow of viscous, incompressible and electrically conducting fluid formed by two cylinders of infinite length with radius a and b such that $a < b$ under the influence of transverse magnetic field. The motion as well as the temperature of the inner cylinder is ramped while the motion together with the temperature of the outer cylinder is fixed. The z - axis is taken along the axis of the cylinder in the vertical upward direction and r'-axis is in the radial direction. A magnetic field of strength B_0 is assumed to be uniformly applied in the direction perpendicular to the direction of flow. In the present physical situation, a constant isothermal heating of $T_w$is applied at the outer surface of the inner cylinder such that $T_w > T_0$. When the time is greater than zero that is $t' > 0$, the temperature of the cylinder is increased or decreased to $T'_0 + \left(T'_w - T_0\right)\frac{t'}{t_0}$, and it begins to move with a velocity proportional to$f(t')$ when $t' \leq t_0$ and thereafter $t' > t_0$ is maintained at constant temperature T_w as presented in **Figure 1**.

The momentum and energy equations governing the present physical situation are given by;

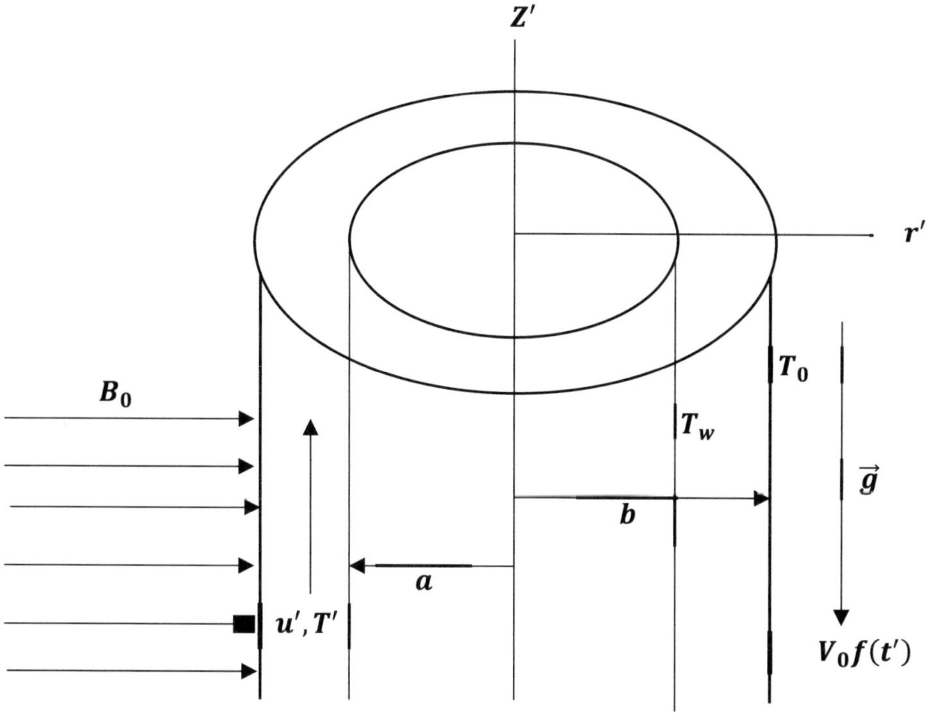

Figure 1.
Physical configuration.

$$\frac{\partial u'}{\partial t'} = \nu\left[\frac{\partial^2 u'}{\partial r'^2} + \frac{1}{r'}\frac{\partial u'}{\partial r'}\right] - \frac{\sigma {B_0}^2 u'}{\rho} + g\beta(T' - T_0) \tag{1}$$

This is valid when the magnetic lines of force are fixed relative to the fluid. If the magnetic field is also having ramped motion with the same velocity as the moving cylinder, the relative motion must be accounted for. In this case the Eq. (1) above is replaced by:

$$\frac{\partial u'}{\partial t'} = \nu\left[\frac{\partial^2 u'}{\partial r'^2} + \frac{1}{r'}\frac{\partial u'}{\partial r'}\right] - \frac{\sigma {B_0}^2}{\rho}(u' - V_0 f(t')) + g\beta(T' - T_0) \tag{2}$$

This is valid when the magnetic lines of force are fixed relative to the moving cylinder. Eqs. (1) and (2) can be combined to obtain the momentum and energy equation respectively.

$$\frac{\partial u'}{\partial t'} = \nu\left[\frac{\partial^2 u'}{\partial r'^2} + \frac{1}{r'}\frac{\partial u'}{\partial r'}\right] - \frac{\sigma {B_0}^2}{\rho}(u' - Kf(t')) + g\beta(T' - T_0) \tag{3}$$

$$\frac{\partial T'}{\partial t'} = \frac{k}{\rho c_p}\left[\frac{\partial^2 T'}{\partial r'^2} + \frac{1}{r'}\frac{\partial T'}{\partial r'}\right] + \frac{Q}{\rho c_p} \tag{4}$$

The relevant dimensional boundary conditions are;

$$t' \le 0 \qquad u' = 0, \quad T' = T_0 \quad \text{For } a \le r' \le b$$

$$t' > 0 \qquad \begin{cases} u' = U_0 f(t'), \quad T' = T_0' + (T_w' - T_0)f(t') & \text{at } r' = a \\ u' = 0, \quad T' = T_w & \text{at } r' = b \end{cases}$$

Where,

$$K = \begin{cases} 0 \text{ if the magnetic field is fixed relative to the fluid} \\ 0.5 \text{ if the velocity of the magnetic field is less than the velocity of the moving cylinder} \\ 1 \text{ if the magnetic field is fixed relative to the moving cylinder} \end{cases}$$

Anand and Kumar [10].

2.1 Non-dimensionalization

The following non-dimensional parameters are defined as:

$$t = \frac{t'\nu}{a^2},\ t_0 = \frac{a^2}{\nu},\ R = \frac{r'}{a},\ \lambda = \frac{b}{a},\ M^2 = \frac{\sigma {B_0}^2 a^2}{\rho\nu},\ \theta = \frac{(T' - T_0)}{(T_w - T_0)},\ Pr = \frac{\mu c_p}{k}$$

$$U = \frac{u'}{v_0},\ K = \frac{V_0}{U_0},\ U_0 = \frac{g\beta(T_w - T_0)a^2}{v},\ Gr = \frac{g\beta(T_w - T_0)a^3}{v^2},\ A = \frac{Q_0 V t_0}{k} \tag{5}$$

Where θ is the dimensionless temperature; Pr is the Prandtl number; M is the Hartmann number and t is the dimensionless time.

Using the non-dimensional parameters in Eq. (5) above, the governing equations of momentum (3) and energy (4) can be written in dimensionless form as:

$$\frac{\partial U}{\partial t} = \left[\frac{\partial^2 U}{\partial R^2} + \frac{1}{R}\frac{\partial U}{\partial R}\right] - M^2(U - K(f(t)) + Gr\theta \tag{6}$$

$$\frac{\partial \theta}{\partial t} = \frac{1}{Pr}\left[\frac{\partial^2 \theta}{\partial R^2} + \frac{1}{R}\frac{\partial \theta}{\partial R}\right] - A \tag{7}$$

The initial conditions for velocity and temperature field in dimensionless form are:

$$t \le 0 \qquad U = 0, \quad \theta = 0 \ \text{ For } \ 1 \le R \le \lambda \tag{8}$$

While the boundary conditions in dimensionless form is given as:

$$t > 0 \qquad \begin{cases} U = f(t), & \theta = f(t) \ \text{ at } R = 1 \\ U = 0, & \theta = 0 \quad \text{ at } R = \lambda \end{cases} \tag{9}$$

Where $\lambda = \frac{b}{a} > 1$

$$f(t) = \begin{cases} \dfrac{t}{t_0} & if\ 0 \le t \le t_0 \\ 1 & if\ t \ge t_0 \end{cases}$$

$$f(t) = H(t)\left(\frac{t}{t_0}\right) - \left(\frac{1}{t_0}\right)(t - t_0)H(t - t_0)$$

Where $H(t)$ is the Heaviside unit step function defined by $H(t) = \begin{cases} 0, t < 0 \\ 1, t \ge 0 \end{cases}$

2.2 Laplace transform

Introducing the Laplace transform on the dimensionless velocity and temperature

$$\overline{U}(R,s)=\int_0^\infty U(R,t)\exp(-st)dt \tag{10}$$

$$\overline{\theta}(R,s)=\int_0^\infty \theta(R,t)\exp(-st)dt \tag{11}$$

(Where s is the Laplace parameter such that $s>0$) applying the properties of Laplace transform on Eqs. (6) and (7) subject to initial condition (8) gives

$$\left[\frac{d^2\overline{U}}{dR^2}+\frac{1}{R}\frac{d\overline{U}}{dR}\right]-\left(M^2+s\right)\overline{U}-{-Gr\overline{\theta}}-M^2K\overline{f}(s) \tag{12}$$

$$\left[\frac{d^2\overline{\theta}}{dR^2}+\frac{1}{R}\frac{d\overline{\theta}}{dR}\right]-sPr\overline{\theta}-\frac{HPr}{s}=0 \tag{13}$$

The boundary conditions (8) becomes

$$\begin{array}{lll}\overline{U}=\overline{f}(s), & \overline{\theta}=\overline{f}(s) & \text{at } R=1\\ \overline{U}=0, & \overline{\theta}=0 & \text{at } R=\lambda\end{array} \tag{14}$$

2.3 Solution

The set of Bessel ordinary differential Eqs. (12) and (13) with the boundary condition (14) are solved for velocity and temperature in the Laplace domain as follows:

$$\begin{aligned}\overline{U}(R,s)&=C_3I_0\left(R\sqrt{M^2+s}\right)+C_4K_0\left(R\sqrt{M^2+s}\right)\\&-\left[\frac{Gr\left(C_1I_0\left(R\sqrt{sPr}\right)+C_2K_0\left(R\sqrt{sPr}\right)\right)}{sPr-\left(M^2+s\right)}\right]+\frac{M^2K\overline{f}(s)}{M^2+s}-\frac{GrH}{\left(M^2+s\right)s^2}\end{aligned} \tag{15}$$

$$\overline{\theta}(R,s)=C_1I_0\left(R\sqrt{sPr}\right)+C_2K_0\left(R\sqrt{sPr}\right)-\frac{A}{s^2} \tag{16}$$

Where;

$$C_1=\frac{s^2\overline{f}(s)K_0\left(\lambda\sqrt{sPr}\right)+A\left[K_0\left(\lambda\sqrt{sPr}\right)-K_0\left(\sqrt{sPr}\right)\right]}{s^2\left[I_0\left(\sqrt{sPr}\right)K_0\left(\lambda\sqrt{sPr}\right)-I_0\left(\lambda\sqrt{sPr}\right)K_0\left(\sqrt{sPr}\right)\right]}$$

$$C_2=\frac{s^2\overline{f}(s)I_0\left(\lambda\sqrt{sPr}\right)+A\left[I_0\left(\lambda\sqrt{sPr}\right)-I_0\left(\sqrt{sPr}\right)\right]}{s^2\left[K_0\left(\sqrt{sPr}\right)I_0\left(\lambda\sqrt{sPr}\right)-K_0\left(\lambda\sqrt{sPr}\right)I_0\left(\sqrt{sPr}\right)\right]}$$

$$C_3=\frac{\overline{f}(s)K_0(\lambda\delta)-A_1[K_0(\lambda\delta)-K_0(\delta)]+A_4[K_0(\lambda\delta)-K_0(\delta)]+[A_2K_0(\lambda\delta)-A_3K_0(\delta)]}{[I_0(\delta)K_0(\lambda\delta)-I_0(\lambda\delta)K_0(\delta)]}$$

$$C_4=\frac{\overline{f}(s)I_0(\lambda\delta)-A_1[I_0(\lambda\delta)-I_0(\delta)]+A_4[I_0(\lambda\delta)-I_0(\delta)]+[A_2I_0(\lambda\delta)-A_3I_0(\delta)]}{[K_0(\delta)I_0(\lambda\delta)-K_0(\lambda\delta)I_0(\delta)]}$$

Where: $= \sqrt{M^2+s}$, $A_1 = \frac{M^2K\bar{f}(s)}{M^2+s}$, $A_4 = \frac{GrA}{(M^2+s)s^2}$, $A_2 = \frac{Gr\left[C_1I_0\left(\sqrt{sPr}\right)+C_2K_0\left(\sqrt{sPr}\right)\right]}{sPr-\left(M^2+s\right)}$

$$A_3 = \frac{Gr\left[C_1I_0\left(\lambda\sqrt{sPr}\right)+C_2K_0\left(\lambda\sqrt{sPr}\right)\right]}{sPr-\left(M^2+s\right)}$$

2.4 Skin friction

The skin – friction is the measure of the frictional force between the fluid and the surface of the cylinder. $(\bar{\tau}_1)$ is the skin-friction at the outer surface of the inner cylinder and $(\bar{\tau}_\lambda)$ is the skin- friction at the inner surface of the outer cylinder. These are obtained by taking the first derivative of the velocity $\overline{U}(R,s)$ given in Eq. (15) with respect to R as follows:

$$\bar{\tau}_1 = \left.\frac{d\overline{U}}{dR}\right|_{R=1} = \sqrt{M^2+s}\left(C_3I_1\left(\sqrt{M^2+s}\right) - C_4K_1\left(\sqrt{M^2+s}\right)\right) - \frac{Gr\sqrt{sPr}}{\left[sPr-\left(M^2+s\right)\right]}\left(C_1I_1\left(\sqrt{sPr}\right) - C_2K_1\left(\sqrt{sPr}\right)\right) \tag{17}$$

$$\bar{\tau}_\lambda = \left.\frac{d\overline{U}}{dR}\right|_{R=\lambda} = \sqrt{M^2+s}\left(C_3I_1\left(\lambda\sqrt{M^2+s}\right) - C_4K_1\left(\lambda\sqrt{M^2+s}\right)\right) - \frac{Gr\sqrt{sPr}}{\left[sPr-\left(M^2+s\right)\right]}\left(C_1I_1\left(\lambda\sqrt{sPr}\right) - C_2K_1\left(\lambda\sqrt{sPr}\right)\right) \tag{18}$$

2.5 Nusselt number

The expression for Nusselt number which is the measure of heat transfer rate on the cylinder is presented in the following form. $Nu_\alpha = \left.\frac{d\bar{\theta}}{dR}\right|_{R=\alpha}$

$$Nu_1 = \left.\frac{d\bar{\theta}}{dR}\right|_{R=1} = \sqrt{sPr}\left(C_1I_1\left(\sqrt{sPr}\right) - C_2K_1\left(\sqrt{sPr}\right)\right) \tag{19}$$

$$Nu_\lambda = \left.\frac{d\bar{\theta}}{dR}\right|_{R=\lambda} = \sqrt{sPr}\left(C_1I_1\left(\lambda\sqrt{sPr}\right) - C_2K_1\left(\lambda\sqrt{sPr}\right)\right) \tag{20}$$

2.6 Mass flow rate

Mass flow rate evaluates the rate of fluid flow through the annulus. It is achieved by taking a definite integral of Eq. (15) with respect to R as shown below:

$$\begin{aligned}\overline{Q} = 2\pi\int_1^\lambda R\overline{U}(R,s)dR &= 2\pi\frac{C_3}{\sqrt{M^2+s}}\left(\lambda I_1\left(\lambda\sqrt{M^2+s}\right) - I_1\left(\sqrt{M^2+s}\right)\right)\\ &- \frac{C_4}{\sqrt{M^2+s}}\left(\lambda K_1\left(\lambda\sqrt{M^2+s}\right) - K_1\left(\sqrt{M^2+s}\right)\right) - \frac{GrC_1}{\sqrt{sPr}\left[SPr-\left(M^2+s\right)\right]}\left(\lambda I_1\left(\lambda\sqrt{sPr}\right) - I_1\left(\sqrt{sPr}\right)\right)\\ &+ \frac{GrC_2}{\sqrt{sPr}\left[sPr-\left(M^2+s\right)\right]}\left(\lambda K_1\left(\lambda\sqrt{sPr}\right) - K_1\left(\sqrt{sPr}\right)\right)\\ &+ \left(\frac{\lambda^2-1}{2}\right)\left(\frac{M^2K\bar{f}(s)}{M^2+s}\right) - \left(\frac{\lambda^2-1}{2}\right)\left(\frac{GrA}{(M^2+s)s^2}\right)\end{aligned} \tag{21}$$

2.7 Riemann sum approximation

Eqs. (15) and (16) are to be inverted in order to determine the velocity and temperature in time domain. Since these equations are difficult to invert in closed form. We use a numerical procedure used in Jha and Apere [14] which is based on the Riemann-sum approximation. In this method, any function in the Laplace domain can be inverted to the time domain as follows.

$$U(R,t) = \frac{e^{\varepsilon t}}{t}\left[\frac{1}{2}\overline{U}(R,\varepsilon) + Re\sum_{n=1}^{M}\overline{U}\left(R,\varepsilon+\frac{in\pi}{t}\right)(-1)^n\right], 1\leq R\leq\lambda \tag{22}$$

where Re refers to the real part of $i = \sqrt{-1}$ the imaginary number. M is the number of terms used in the Riemann-sum approximation and ε is the real part of the Bromwich contour that is used in inverting Laplace transforms. The Riemann-sum approximation for the Laplace inversion involves a single summation for the numerical process, its accuracy depends on the value of ε and the truncation error dictated by M. According to Tzou [29], the value of εt that best satisfied the result is 4.7.

2.8 Validation of results

In order to validate the results obtained from the Riemann sum approximation methods we use the partial differential equation parabolic and elliptic (PDEPE) method and compared the result. In General, it is given in the form:

$$c\left(x,t,u,\frac{\partial u}{\partial t}\right)\frac{\partial u}{\partial t} = x^{-m}\frac{\partial}{\partial x}\left(x^m f\left(x,t,u,\frac{\partial u}{\partial x}\right)\right) + s\left(x,t,u,\frac{\partial u}{\partial x}\right) \tag{23}$$

Initial condition $U(x,t_0) = U_0 x$ Boundary conditions- one at each boundary $(x,t,u) + q(x,t)f\left(x,t,u,\frac{\partial u}{\partial x}\right) = 0$. These comparisons are analyzed on the tables.

2.9 Result and discussion

A MATLAB program is written in order to depict the effect of the flow parameters such as the Hartman number (M), Prandlt number (Pr), Grashoff number (Gr), Heat source/sink parameter (A) and ratio of radii (λ) on Velocity (U), Temperature (T), Nusselt number at the outer surface of the inner cylinder (Nu_1), Nusselt number at the inner surface of the outer cylinder (Nu_λ), Skin friction (λ) at the outer surface of the inner cylinder (τ_1), Skin friction at the inner surface of the outer cylinder (τ_λ) and mass flow rate (Q).

Figure 2 illustrated the temperature profile for different values of time. It is seen from this graph that the temperature increases with increase in time. Increase in radii ratio lead to an increase in temperature as show in **Figure 3**. As the heat generating or absorbing parameter increase, a decrease in temperature is noticed as depicted in **Figure 4**.

Figure 5 illustrated the effect of time on fluid velocity for cases $(K = 0.0$ if the magnetic field is fixed relative to the fluid), $(K = 0.5$ if the velocity of the magnetic field is less than the velocity of the moving cylinder) and $(K = 1.0$ if the magnetic field is fixed relative to the moving cylinder) and for Prandtl number $(Pr = 0.71\ Air\ and\ Pr = 7.0\ Water)$. These graphs show that the fluid velocity

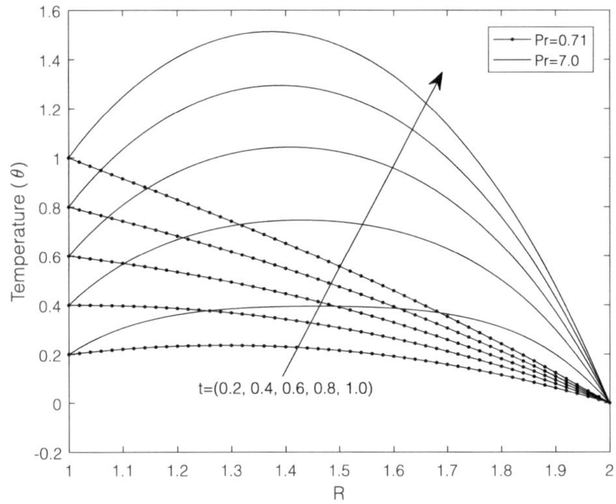

Figure 2.
Temperature distribution for different values of time $(t)(\lambda = 2.0, A = -2.0)$.

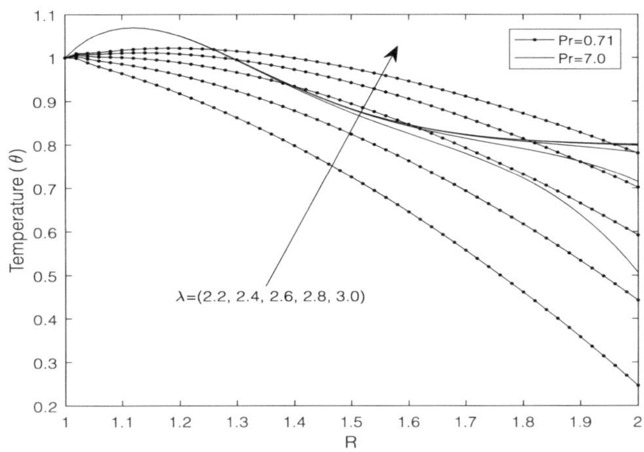

Figure 3.
Temperature distribution for different values of radii ratio (λ) $(t = 0.4, A = -2.0)$.

increase as time increases for all cases. It is interesting to note that, $Pr = 0.71$ converges faster than $Pr = 7.0$. Also, the value of velocity becomes higher as the value of K increases. **Figure 6** depicted the influence of Hartmann number on the fluid velocity for cases $(K = 0.0, K = 0.5\ and\ K = 1.0)$ and $Pr = 7.0$. the fluid velocity decreases for cases $K = 0.0\ and\ K = 0.5$ This is physically true because application of magnetic field to an electrically conducting fluid give rise to resistivity force which is known as Lorentz force, this force has the tendency to decelerate fluid flow in the boundary layer region. Hartman number increase the fluid velocity for case $K = 1.0$. This implies that the magnetic field is supporting the fluid motion. Although a flow reversal is noticed for $= 7.0\ at R > 1.4$. **Figure 7** show the outcome of radii ratio on fluid velocity for cases $(K = 0.0, K = 0.5\ and\ K = 1.0)$ respectively and for $Pr = 7.0$. It is evident from the graph that, increase in radii ratio result to an increase in fluid velocity for all cases of K considered. The value of velocity becomes higher as the value of K increases. Heat generating/absorbing parameter (A) has a decreasing influence on the fluid velocity for all cases considered

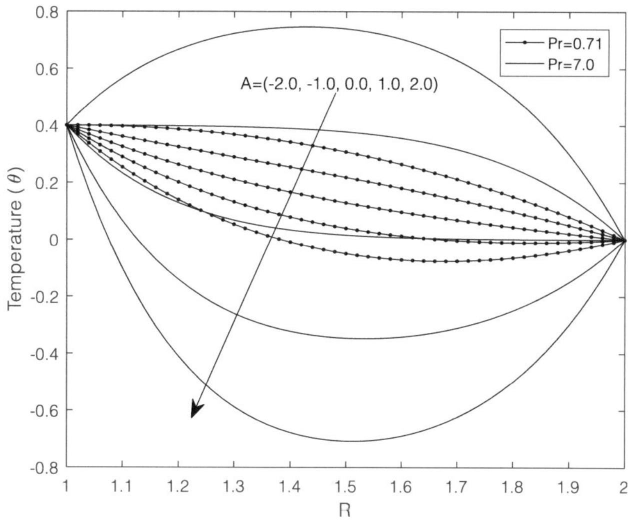

Figure 4.
Temperature profile for different values of heat source/sink (A) ($t = 0.4, \lambda = 2.0$).

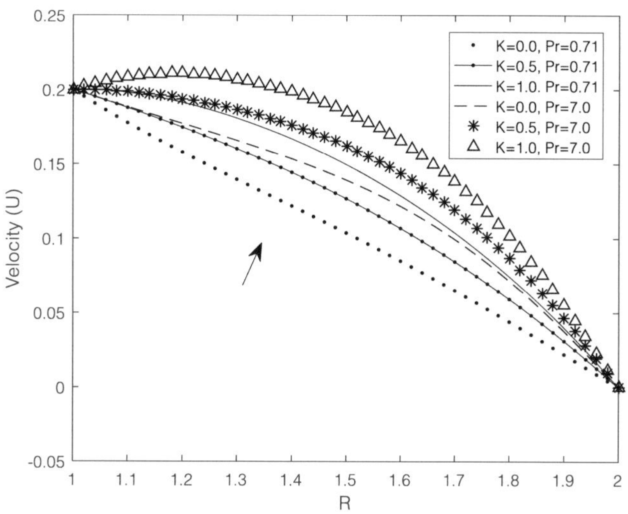

Figure 5.
Velocity distribution for different values of time (t and K)($Pr = 7.0, \lambda = 2.0, M = 2.0$ $Gr = 5.0, A = -2.0$).

($K = 0.0, K = 0.5$ *and* $K = 1.0$) as illustrated in **Figure 8**. **Figure 9** presents the effect of Grashof number on fluid velocity for cases ($K = 0.0, K = 0.5$ *and* $K = 1.0$) and for $Pr = 7.0$. It is evident from the graph that increase in thermal buoyancy force lead to an increase in the fluid velocity for all cases of K.

Figure 10 presents the effect of Hartmann number on skin friction at the outer surface of the inner cylinder(τ_1) for cases ($K = 0.0, K = 0.5$ *and* $K = 1.0$) and for($Pr = 7.0$). It is evident from the graph that, the Hartmann number decrease the skin friction (τ_1) for all cases of K. **Figure 11** describe the impact of radii ratio on the skin friction (τ_1) for cases ($K = 0.0, K = 0.5$ *and* $K = 1.0$) and for($Pr = 7.0$). The graph show that, increase in radii ratio result to an increase in skin friction (τ_1). **Figure 12** depicts the influence of heat generating/absorbing parameter (A) for for cases ($K = 0.0, K = 0.5$ *and* $K = 1.0$) and for ($Pr = 7.0$). It is noticed from these

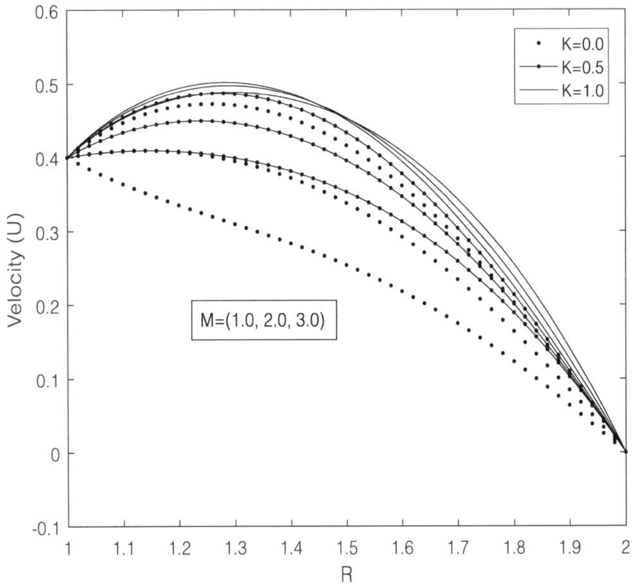

Figure 6.
Velocity distribution for different values of Hartmann number $(M\ and\ K)$ $(Pr = 7.0, \lambda = 2.0, t = 0.4$ $Gr = 5.0, A = -2.0)$.

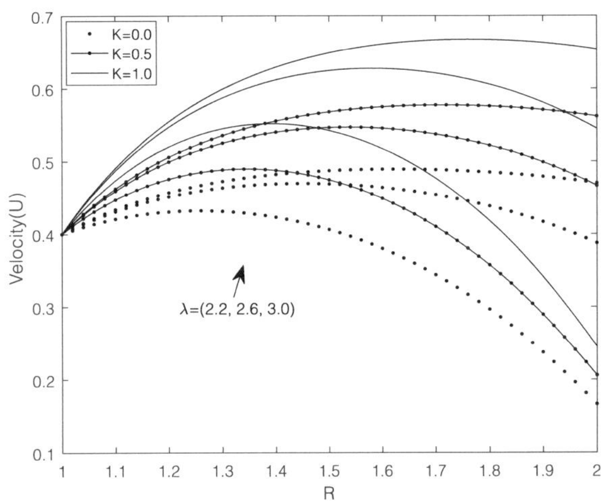

Figure 7.
Velocity distribution for different values of radii ratio $(\lambda\ and\ K)$ $(Pr = 7.0, A = -2.0, t = 0.4\ Gr = 5.0, M = 2.0)$.

graphs that, Heat absorption has a retarding effect on the skin friction (τ_1) for all cases of K. It is essential to note that, a reverse flow occur on $Pr = 7.0$ at $t = 0.8$ for all cases of K considered. **Figure 13** illustrate the effect of Grashof number on skin friction (τ_1) for cases $(K = 0.0, K = 0.5\ and\ K = 1.0)$ and for $(Pr = 7.0)$. The thermal buoyance force is seen to increase the skin friction (τ_1) from the graph. **Figure 14** depict the effect of Hartmann number on skin friction at the inner surface of the outer cylinder (τ_λ) for cases $(K = 0.0, K = 0.5\ and\ K = 1.0)$ and for $(Pr = 7.0)$. It is seen from the graph that, Hartmann number increase the skin friction(τ_λ) for case $K = 0.0$ while the opposite effect is noticed for both case

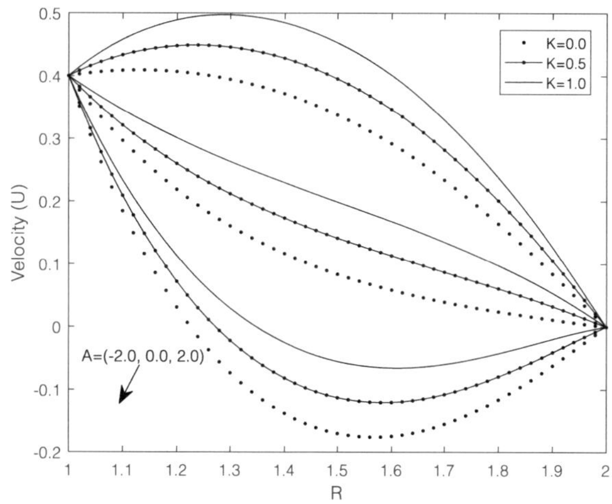

Figure 8.
Velocity distribution for different values of heat source/sink $(A\ and\ K)$ $(Pr = 7.0, \lambda = 2.0, t = 0.4$ $Gr = 5.0, M = 2.0)$.

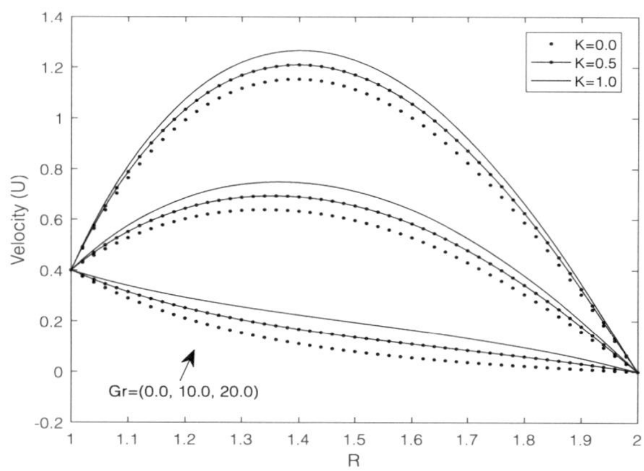

Figure 9.
Velocity distribution for different values of Grashof number $(Gr\ and\ K)$ $(Pr = 7.0, \lambda = 2.0, t = 0.4,$ $A = -2.0, M = 2.0)$.

$K = 0.5\ and\ K = 1.0$. **Figure 15** show the impact of radii ratio on the skin friction (τ_λ) for cases $(K = 0.0, K = 0.5\ and\ K = 1.0)$ and for $(Pr = 7.0)$. The graph show that, increase in radii ratio lead to a decrease in skin friction (τ_λ). **Figure 16** present the influence of heat generating/absorbing parameter (A) for cases $(K = 0.0, K = 0.5\ and\ K = 1.0)$ and for $(Pr = 7.0)$. It is evident from the graph that, there is an enhancement in skin friction(τ_λ) as the heat generating/absorbing parameter increase for all cases of K. **Figure 17** demonstrate the effect of Grashof number on skin friction (τ_λ) for cases $(K = 0.0, K = 0.5\ and\ K = 1.0)$ and for $(Pr = 7.0)$. The thermal buoyance force is seen to decrease the skin friction (τ_λ) from the graph.

Figure 18 illustrate the effect of Hartmann number on mass flow rate (Q) for cases $(K = 0.0, K = 0.5\ and\ K = 1.0)$ and for $(Pr = 7.0)$. The Hartman number decreases the volume flow rate. **Figure 19** show the impact of radii ratio on mass flow rate (Q) for cases $(K = 0.0, K = 0.5\ and\ K = 1.0)$ and for $(Pr = 7.0)$. It is seen

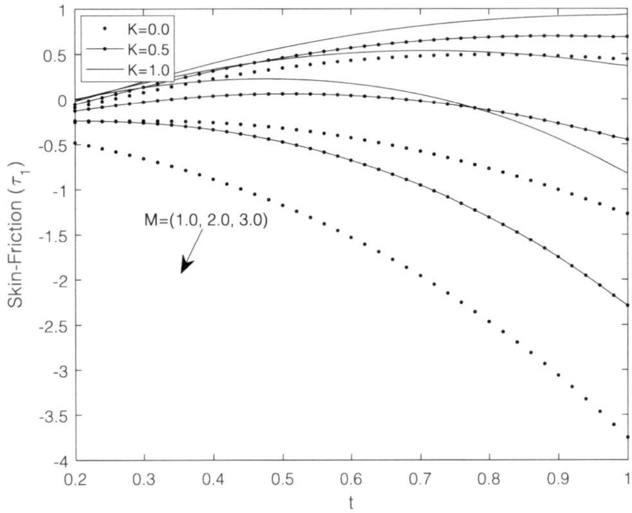

Figure 10.
Variation of skin friction (τ_1) for different values of Hartmann number (M and K)($Pr = 7.0, A = -2.0,$ $\lambda = 2.0$ $Gr = 5.0$).

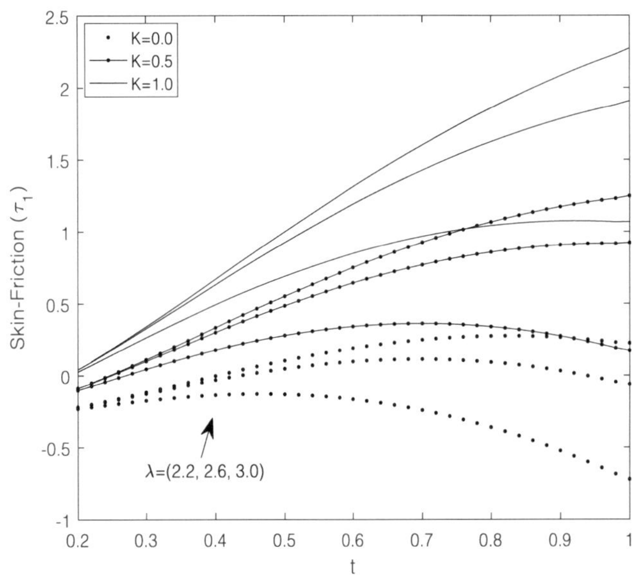

Figure 11.
Variation of skin friction (τ_1) for different values of radii ratio (λ and K) ($Pr = 7.0, A = -2.0, M = 2.0$ $Gr = 5.0$).

from the graph that, the radii ratio increase the mass flow rate for all cases of K. **Figure 20** present the influence of heat generating/absorbing parameter on mass flow rate (Q) for cases ($K = 0.0, K = 0.5$ *and* $K = 1.0$) and for ($Pr = 7.0$). It is noticed from the graph that, the heat generating/absorbing parameter decrease the mass flow rate for all cases of K. **Figure 21** depict the effect of Grashof number on mass flow rate (Q) for cases ($K = 0.0, K = 0.5$ *and* $K = 1.0$) and for ($Pr = 7.0$). It is evident from the graph that, the thermal buoyancy force increase the mass flow rate for all cases of K.

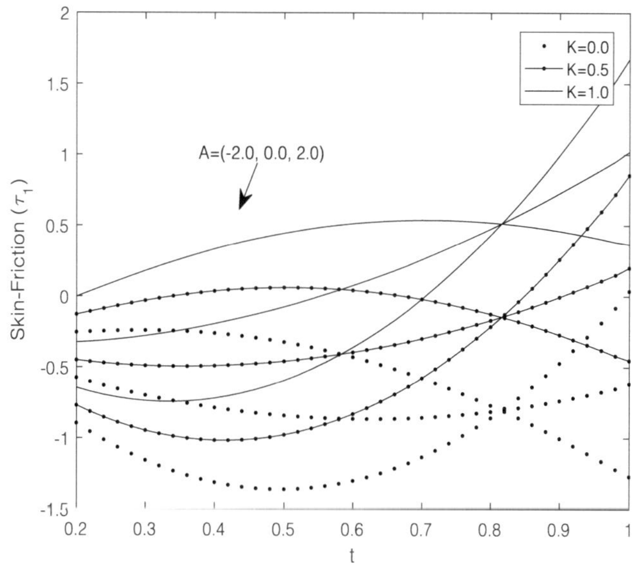

Figure 12.
Variation of skin friction (τ_1) *for different values of heat source/sink* $(A$ *and* $K)$ $(Pr = 7.0, \lambda = 2.0, M = 2.0$ $Gr = 5.0)$.

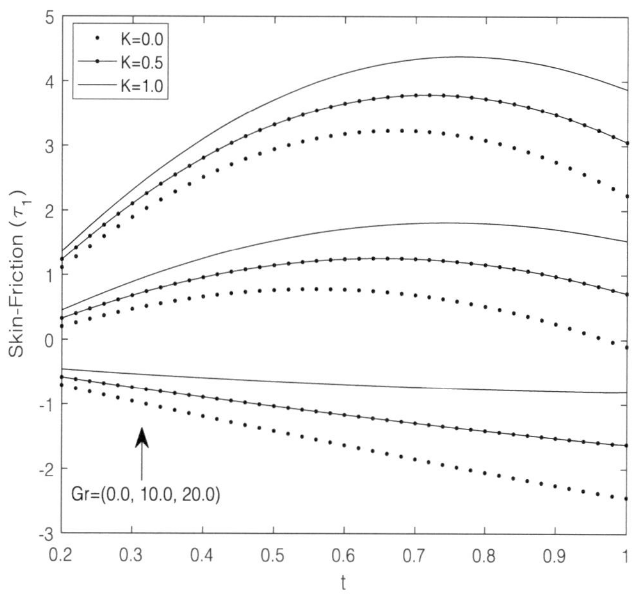

Figure 13.
Variation of skin friction (τ_1) *for different values of Grashof number* $(Gr$ *and* $K)$ $(Pr = 7.0, \lambda = 2.0, M = 2.0$ $Gr = 5.0, A = -2.0)$.

3. Conclusion

The study of MHD natural convection flow of constant heat source/sink in an annulus due to ramped motion and ramped temperature of the boundaries have been carried out. The Laplace transform techniques have been used and the time domain solution was obtained using the Riemann sum approximation. The effect of the governing parameters such as the Hartmann number (M), radii ratio (λ), time (t), Grashof number (Gr) Heat generating/absorbing parameter (A) and

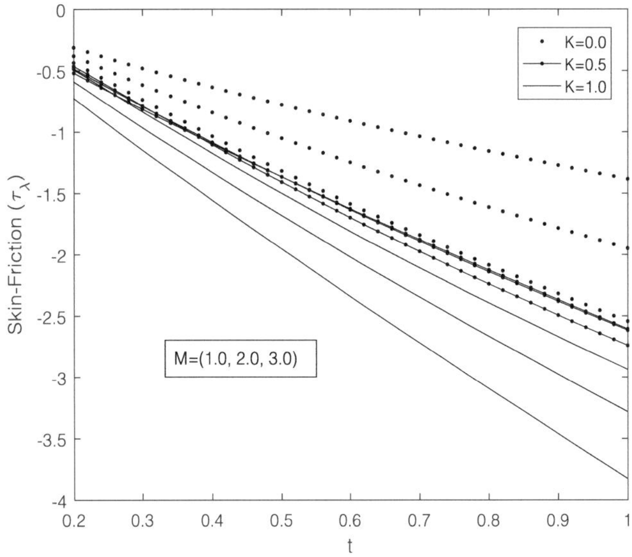

Figure 14.
Variation of skin friction (τ_λ) *for different values of Hartmann number* $(M$ *and* $K)(Pr = 7.0, \lambda = 2.0,$ $A = -2.0$ $Gr = 5.0)$.

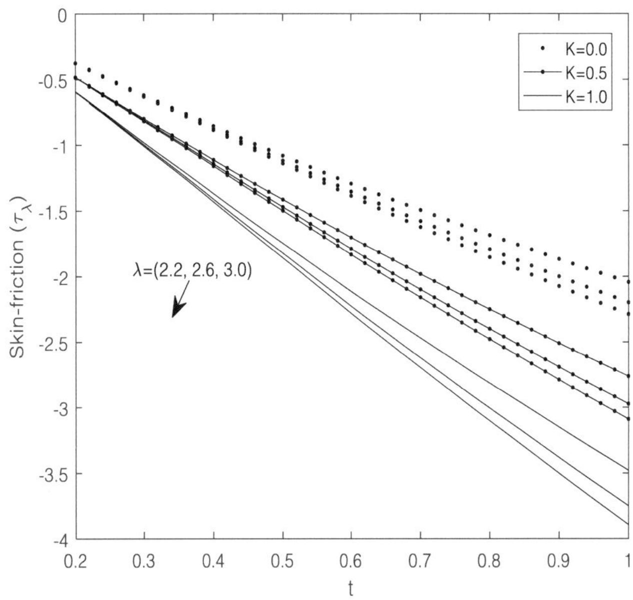

Figure 15.
Variation of skin friction (τ_λ) *for different values of radii ratio* $(\lambda$ *and* $K)$ $(Pr = 7.0, M = 2.0, A = -2.0$ $Gr = 5.0)$.

Prandtl number water $(Pr = 7.0)$ on the dimensionless fluid velocity (U), temperature (θ), mass flow rate (Q) and skin – friction $(\tau_1$ *and* $\tau_\lambda)$ at both surfaces of the cylinder considering three cases of the velocity of the magnetic field ($K = 0.0$ when the magnetic field is fixed relative to the fluid, $K = 0.5$ when the velocity of the magnetic field is less than the velocity of the moving cylinder and $K = 1.0$ when the magnetic field is fixed relative to the moving cylinder) have been analyzed with the help of line graphs.

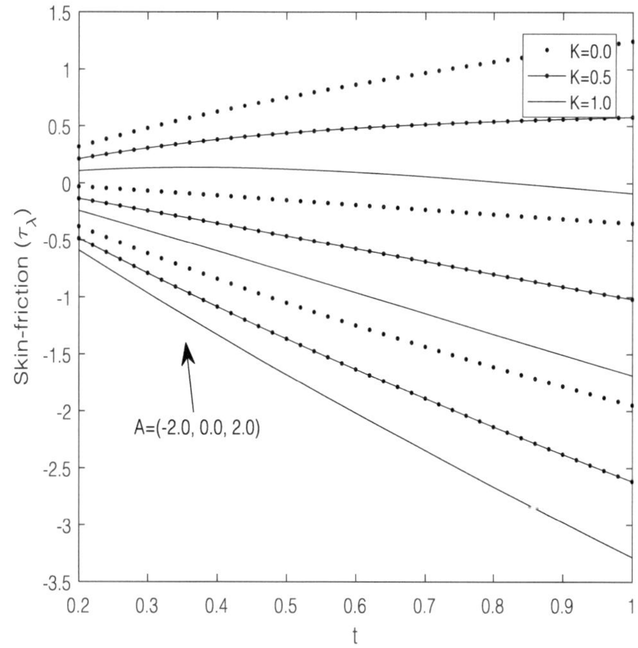

Figure 16.
Variation of skin friction (τ_λ) *for different values of heat source/sink* $(A\ and\ K)(Pr = 7.0, M = 2.0, \lambda = 2.0$ $Gr = 5.0)$.

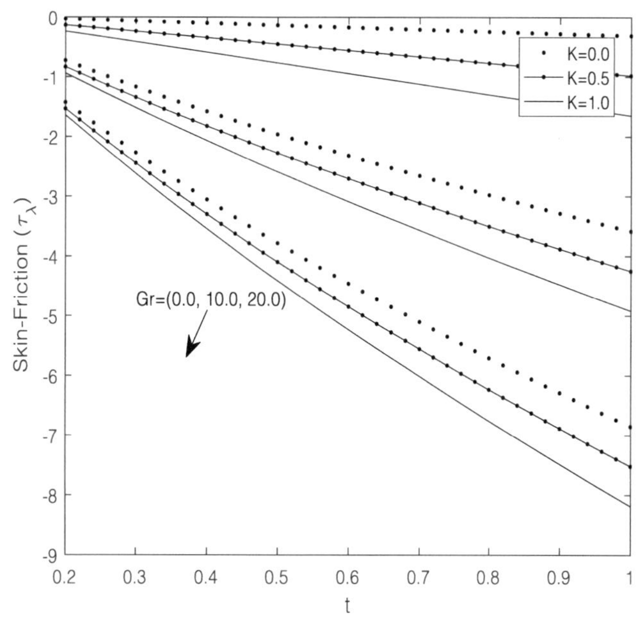

Figure 17.
Variation of skin friction (τ_λ) *for different values of Grashof number* $(Gr\ and\ K)(K = 0.0, M = 2.0, \lambda = 2.0$ $A = -2.0)$.

The noteworthy conclusions are summarized as follows:

- Hartmann number has a retarding effect on the skin friction (τ_1) and mass flow rate. It also decreases fluid velocity for cases $(K = 0.0\ and\ K = 0.5)$ the reverse

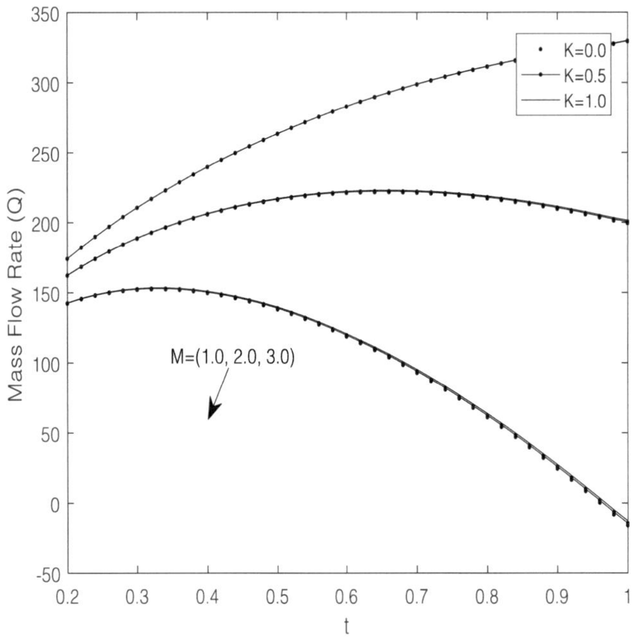

Figure 18.
Variation of mass flow rate (Q) *for different values of Hartmann number* $(M \text{ and } K)(K = 0.0, \lambda = 2.0, A = -2.0, Gr = 5.0)$.

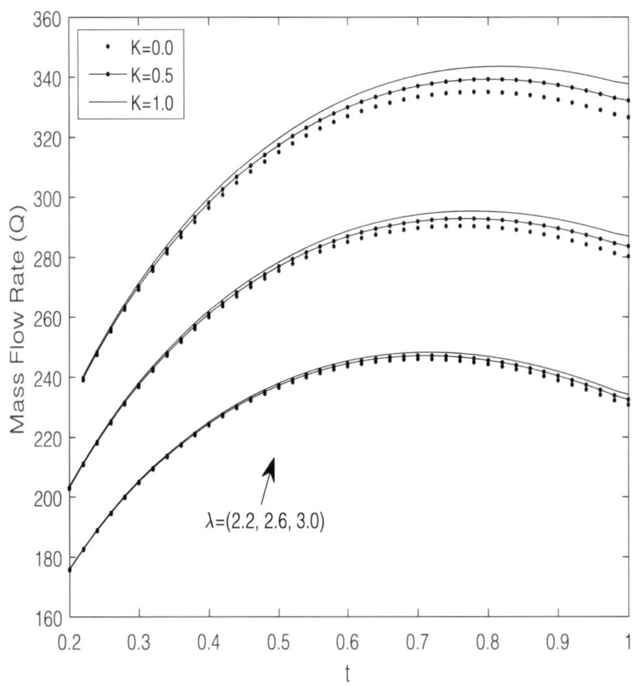

Figure 19.
Variation of mass flow rate (Q) *for different values of radii ratio* $(\lambda \text{ and } K)(K = 0.0, M = 2.0, A = -2.0, Gr = 5.0)$.

effect is noticed for case $(K = 1.0)$. increase in Hartmann number lead to an increase in skin friction (τ_λ) for case $(K = 0.0)$ but decreases it for cases $(K = 0.0 \textit{ and } K = 0.5)$

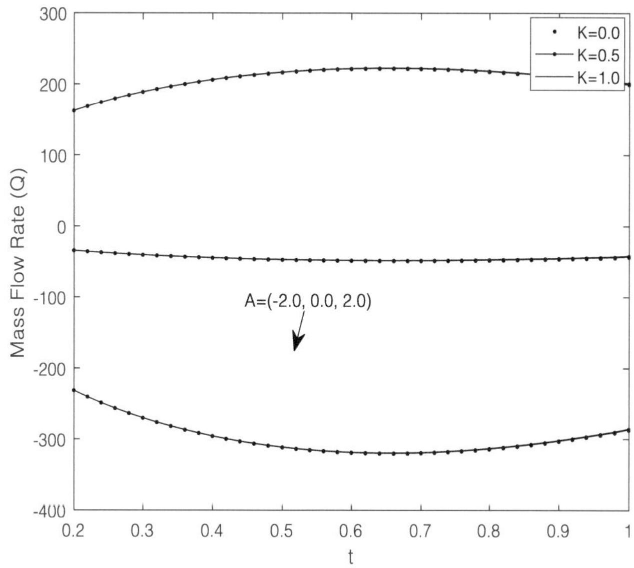

Figure 20.
Variation of mass flow rate (Q) *for different values of heat source/sink* $(A$ *and* $K)(K = 0.0, M = 2.0$ $\lambda = 2.0, Gr = 5.0)$.

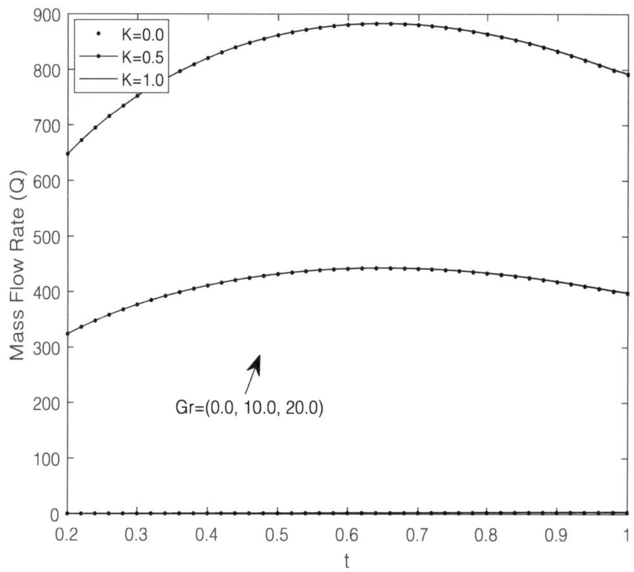

Figure 21.
Variation of mass flow rate (Q) *for different values of Grashof number* $(Gr$ *and* $K)(K = 0.0, M = 2.0,$ $\lambda = 2.0, A = -2.0)$.

- Thermal buoyancy force and radii ratio increase mass flow rate when = 7.0. As the radii ratio and Grashof number increase, there is an increase in fluid velocity and skin friction (τ_1) for all cases of K and for $Pr = 7.0$. The reverse effect occurs for skin friction (τ_λ). Radii ratio also increases the fluid temperature

- Heat generating/absorbing parameter has a retarding effect on fluid velocity, temperature and skin friction (τ_1) while it enhances the skin friction (τ_λ) for all cases of K and for $Pr = 7.0$. It decreases mass flow rate when $Pr = 7.0$.

Nomenclature

t'	Dimensional time (s)
u'	Velocity (m/s)
r'	Dimensional radial coordinate
U	Dimensionless velocity (m/s)
B_0	Constant magnetic flux density
T_0	Reference temperature (K)
I_n	Modified Bessel's function of first kind of order n
K_n	Modified Bessel's function of second kind of order n
T_w	Constant temperature (K)
M	Hartmann number
Pr	Prandtl number
Nu	Nusselt number
Gr	Grashof number
Q	Dimensionless volume flow rate $(kgs^{-1}m^{-2})$
a	Radius of the inner cylinder
b	Radius of the outer cylinder
t	Dimensionless time (s)
R	Dimensionless radial coordinate
g	Gravitational acceleration (m/s^2)
c_p	Specific heat at constant pressure $(J/kg/K)$

Greek letters

υ	Fluid kinematic viscosity (m^2/s)
τ	(Skin-friction)
ρ	Density (kg/m^3)
β	Coefficient of thermal expansion (K^{-1})
λ	Ratio of radii $\left(\frac{b}{a}\right)$
σ	Electrical conductivity of the fluid $(W/m.K)$
α	Thermal diffusivity (m^2/s)
μ	Magnetic diffusivity (m^2/s)

Author details

Khadijah Lawal* and Haruna Jibril
Ahmadu Bello University Zaria, Kaduna State, Nigeria

*Address all correspondence to: khadijahlawal19@gmail.com

References

[1] Chandran P, Sacheti NC, Singh AK. Natural convection near a vertical plate with ramped wall temperature. Heat and Mass Transfer. 2005;**41**:459-464

[2] Jha BK, Samaila AK, Ajibade AO. Natural convection flow of heat generating/bbsorbing fluid near a vertical plate with ramped temperature. JEAS. 2012;**2**:61-68

[3] Seth GS, Ansari MS, Nandkeolyar R. MHD unsteady hydro-magnetic Couette flow within a porous channel. Tamkang Journal of Science and Engineering. 2011;**14**:7-14

[4] Nandkeolyar R, Das M. Unsteady MHD free convection flow of a heat absorbing dusty fluid past a flat plate with ramped wall temperature. Afrika Matematika. 2014;**25**:779-798

[5] Seth GS, Ansari MS, Nandkeolyar R. MHD natural convection flow with radiative heat transfer past an impulsively moving plate with ramped wall temperature. Heat and Mass Transfer. 2011;**47**:551-561

[6] Seth G, Hussain SM, Sarkar S. Hydro magnetic natural convection flow with heat and mass transfer of a chemically reacting and heat absorbing fluid past an accelerated moving plate with ramped temperature and ramped surface concentration through a porous medium. Bulgarian Chemical Communications. 2016;**48**:659-670

[7] Lawal KK, Jibril HM. Unsteady MHD natural convection flow of heat generating/absorbing fluid near a vertical plate with ramped temperature and motion. Journal of Taibah University for Science. 2019;**13**(1): 433-439

[8] Lawal KK, Jibril HM. Time dependent MHD natural convection flow of heat generating/absorbing fluid near a vertical porous plate with ramped boundary conditions. Euro Afro Studies International Journal. 2019;**1**(1):81-122

[9] Jha BK, Jibril HM. Hydro magnetic flow due to ramped motion of the boundary. International Journal of Applied Mechanics and Engineering. 2010;**15**:1101-1109

[10] Kumar A, Singh AK. Transient magneto hydrodynamic Couette flow with ramped velocity. International Journal of Fluid Mechanics. 2010;**I37**: 435-446

[11] Jha BK, Jibril H. Effect of transpiration on the MHD flow near a porous plate having ramped motion. Journal of the Physical Society of Japan. 2011;**80**:64684

[12] Jha BK, Jibril H. Time dependent MHD couette flow due to ramped motion of one of the boundaries. International Journal of Applied Mechanics. 2012;**17**(2):383-399

[13] Jha BK, Jibril H. Time dependent MHD Couette due to ramped motion of one the porous plate. International Journal of Applied Mechanics and Engineering. 2013;**18**(4):1039-1056

[14] Jha BK, Apere CA. Unsteady MHD Couette flow in annuli, the Riemann Sum approximation approach. Journal of Physical Society Japan. 2010;**79**: 124403 (5 pp.)

[15] Jha BK, Apere CA. Unsteady MHD two- phase Couette flow of fluid particles suspension in an annulus. AIP Advances. 2011;**1**:042121

[16] Anand. Effect of radial magnetic field on free convective flow over ramped velocity moving vertical cylinder with ramped type temperature and concentration. Fluid Mechanica. 2016;**9**(6):2655-2864

[17] Anand. Effect of radial magnetic field on natural convection flow in alternate conducting vertical concentric annuli with ramped temperature. Engineering Science and Technology. An International Journal. 2016;**19**(3): 1436-1451

[18] Anand. Effect of velocity of applied magnetic field on natural convection over ramped type moving inner cylinder with ramped type temperature. International Journal of Mechanica Sciences. 2017;**131**:625-632

[19] Taiwo YS. Exact solution of an MHD natural convection flow in vertical concentric annulus with heat absorption. International Journal of system engineering. 2017;**1**(1): 15-24

[20] Zakaria MN, Hussanan A, Khan I, et al. The effects of radiation on free convection flow with ramped wall temperature in Brinkman type fluid. Jurnal Teknologi. 2013;**62**:33-39

[21] Jha BK, Apere CA. Magneto hydrodynamic transient free convection flow in a vertical annulus with thermal boundary condition of second kind. International Journal of Heat and Mass Transfer. 2012;**134**(4):042502

[22] Tasawar H et al. Analytic solution for MHD flow in an annulus. Communications in Nonlinear Science and Numerical Simulation. 2010;**15**(5): 1224-1227

[23] Kundu PK, Das K, Acharya N. Flow features of a conducting fluid near an accelerated vertical plate in porous medium with ramped wall temperature. Journal of Mechanics. 2014;**30**:277-288

[24] Samiulhaq, Ahmed S, Vieru D, et al. Unsteady magnetohydrodynamic free convection flow of a second grade fluid in a porous medium with ramped wall temperature. PLoS. 2014;**9**:e88766

[25] Samiulhaq, Khan I, Ali F, et al. Free convection flow of a second-grade fluid with ramped wall temperature. Heat Transfer Research. 2014;**45**:579-588

[26] Seth GS, Nandkeolyar R, Ansari MS. Effect of rotation on unsteady on unsteady hydro magnetic natural convection flow past an impulsively moving vertical plate with thermal diffusion and heat absorption. International Journal of Applied Mathematics and Mechanics. 2011;**7**: 52-69

[27] Lawal KK, Jibril HM. Impact of relative motion of a magnetic field on unsteady magneto-hydrodynamic natural convection flow with a constant heat source/sink. Heat Transfer. 2020;**1**: 1-21. DOI: 10.1002/htj.21888

[28] Lawal KK, Jibril HM. Influence of relative motion of magnetic field on time-dependent MHD natural convection flow. Heat Transfer. 2020;**1**: 1-20. DOI: 10.1002/htj.21890

[29] Tzou DY. Macro to Micro Scale Heat Transfer: The Lagging Behavior. London: Taylor and Francis; 1997

Section 3

Novel Application

Chapter 4

Heat Exchangers for Electronic Equipment Cooling

Abdelhanine Benallou

Abstract

Recent developments in the electronic equipment market have been very demanding on two important design parameters: the size of the equipment and the efficiency of the cooling system. Indeed, the race for more applications handling in reduced sizes in the case of smartphones requires the use of important amounts of energy in tiny volumes. Similar constraints are encountered in the design of the new generation of vehicles (electric cars, hybrid vehicles, high-speed trains, airplanes), which impose the use of highly integrated electronic structures, resulting in significant power densifications (up to several hundred Watts/cm^2). In some CPU boards, the power generated per unit chip area is in the order of 500 kW/m^2. Cooling of such boards requires low volume and lightweight heat exchangers to transfer tremendous amounts of heat. The same situation is encountered for most newly developed demotics' equipment. This chapter reviews available state-of-the-art technologies for electronic equipment cooling, including *passive* and *active techniques*, as well as *one and two-phase heat exchange*. Directions for the design of the different heat exchangers will also be given.

Keywords: passive cooling, active cooling, spray cooling, refrigerated cooling, heat pipes, cold plates, vapor chambers, microchannel heat sinks, thermal resistance, carrying capacity, thermal program, effective thermal conductivity

1. Introduction

Operation of active electronic components (transistors, integrated circuits, microprocessors, CPUs, GPUs, etc.) generates significant amounts of heat, around 500 kW/m^2, quite comparable to the flux densities encountered at the nose of a space shuttle entering the atmosphere.

If this heat is not extracted, it leads to important increases in the temperature of these components. Such increases ineluctably induce degradation of the bonding wires, delamination of solders, and/or the appearance of leakage currents. Moreover, temperature gradients between electronic chips and their soles generate cyclical thermomechanical stresses throughout the lifetime of the chips, which results in thermal fatigue, leading ultimately to component failure.

Thus, one of the problems with the operation of active electronic components is energy dissipation; How can an electronic system get rid of the heat its operation generates? Usually, for each electronic component, manufacturers specify a maximum operating temperature (the maximum junction operating temperature, T_{op}).

If they are not properly cooled when operating, electronic components will certainly reach T_{op} and probably go over it. They then can completely lose their properties, which could lead to an alteration of the operation of the circuit boards or the systems they compose. This alteration can manifest through a complete breakdown of the systems or through a shortening of the mean time between failures (MTBF), leading to a premature aging.

Consequently, design techniques of electronic systems always consider a built-in cooling system, often called a *heatsink* to allow the component to get rid of the generated energy by dissipating it toward its surroundings. Heat sinks are actually special *heat exchangers*; they ensure the evacuation of the heat generated by the operation of electronic components.

This chapter reviews state-of-the-art technologies available for electronic equipment cooling. These technologies include *passive* and *active*, as well as *one and two-phase heat exchange*. Directions for the design of the different heat exchangers are also be given.

2. Passive heatsink technologies

Passive heat sinks are heat exchangers in the form of finned radiators, generally made of aluminum. They are called *passive* because they do not have any moving mechanical component (fan) designed to force airflow. Air moves through them only due to density difference: Hot air goes upward as it is replaced by cooler air. It is for this reason that passive sinks transfer heat to air essentially by natural convection[1]. This type of heat sinks is, by far, preferred for cooling electronic systems [2]; they are cost-effective, simple to find and assemble, and they generate no power consumption or noise.

The most widely used passive heat exchangers are *finned natural convection heat sinks*.

2.1 Finned heat exchangers

In their simplest version, these heat sinks are constituted of finned surfaces. **Figure 1** shows examples of *finned heat exchangers* used as heat sinks on electronic components. They are usually constituted of materials having a high thermal conductivity, such as aluminum or copper. But other new materials, like ceramics, can also be found.

The sink is usually mounted on the electronic component, which generates heat, thus increasing its heat transfer area[2] with ambient air. **Figure 2** shows a transistor mounted on a finned heatsink.

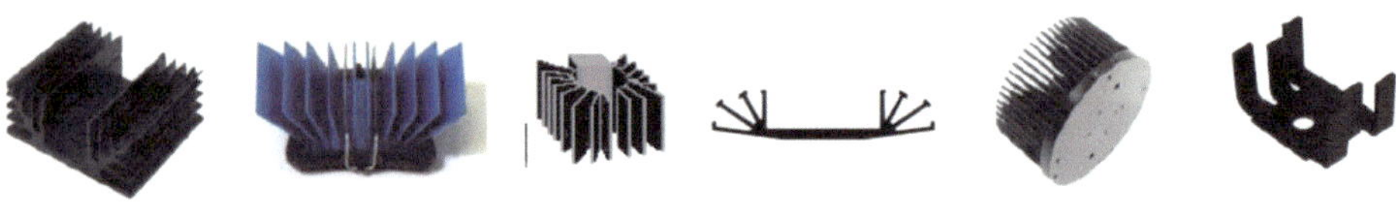

Figure 1.
Examples of finned heat sinks. Sources [3, 4].

[1] For a better understanding of natural convection see Ref. [1].

[2] The heat transfer area is the surface thru which energy is transferred from the electronic component to the environment.

Figure 2.
A heatsink fixed on a transistor. Source [5].

It can be seen from this figure that this type of mounting permits to increase the heat transfer area; the heat generated by the transistor is transferred to ambient, not only through the external surface of the transistor but also through the surfaces of all the fins.

In such situations, the heat transfer process depends on several parameters such as the temperature difference between air and the electronic component, the total area of the fins and their position (vertical/horizontal), the fins spacing, etc.

2.2 Design fundamentals

2.2.1 Thermal resistance

Despite this complexity of the heat transfer process, design techniques usually use a simple model to represent the flow of energy through a heatsink. This simple model is based on a similarity between the way that heat moves from one medium to another and the way that electric current flows from one potential to another. We know that if a point at a potential V_0 (**Figure 3a**) is separated by an electrical resistance R from a second point (at V_1), and then an electric current i flows between these two points, such that:

$$i = \frac{V_0 - V_1}{R} \tag{1}$$

Similarly, it can be shown that the power, Q, of heat dissipation between a point at temperature T_c and a point at temperature T_a (**Figure 3b**) is given by:

$$Q = \frac{T_c - T_a}{R_{th}} \tag{2}$$

Where R_{th} is called the *thermal resistance* between these points [6].

R_{th} is expressed differently for conduction and convection as follows:

- For a conduction heat transfer:

$$R_{th}^{cond} = \frac{L}{kA} \tag{3}$$

Figure 3.
Thermal resistance model for heat flow. (a) Electric current i *flows between* V_o *and* V_1*. (b) Heat flux* Q *flows between* T_c *and* T_a*.*

Where L is the material thickness, k is the thermal conductivity, and A is the heat transfer area.

- For a convection heat transfer:

$$R_{th}^{conv} = \frac{1}{hA} \tag{4}$$

Where h is the convective heat transfer coefficient and A is the heat transfer area.

2.2.2 Electronic component without a heatsink

Consider an electronic component with junction temperature T_j, in an ambient environment (air for example) at T_a.

As shown in **Figure 4**, the heat generated by the component is transferred by conduction from the junction to the external surface of the case. Then, heat is conveyed by convection and radiation to the ambient environment.

For normal operating temperatures, conduction and convection are the prevailing modes. The total heat transfer resistance is therefore given by the following:

$$R_{JA} = R_{JC} + R_{CA} \tag{5}$$

Where:
R_{JA} is the junction to ambient resistance.
R_{JC} is the conduction resistance.
R_{CA} is a convection resistance.

Note that R_{JC} is generally quite low, but R_{CA} is usually high enough to limit heat transfer from the component to environment.

2.2.3 Electronic component with a heatsink

Figure 5 shows the electronic component of **Figure 4** to which a finned heatsink was added in order to take advantage of its large contact area with ambient, thus permitting a better spread of the heat generated by the component.

In this case, as well, heat is transferred from the component to the ambient essentially by conduction and convection. Energy transfers are therefore represented by a series of thermal resistances:

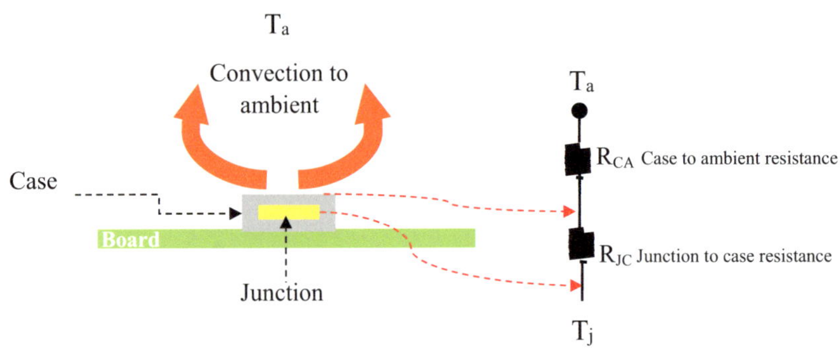

Figure 4.
Heat resistance of a sink.

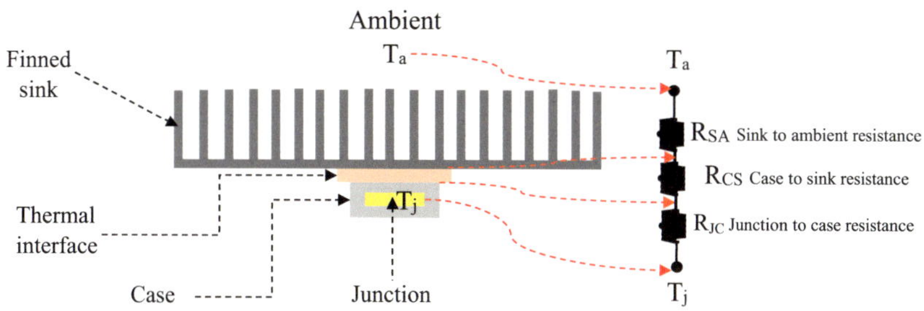

Figure 5.
Total resistance of a heatsink.

R_{JC}, *the junction to case thermal resistance*; a conductive resistance which depends on the thickness, e_{JC} (between the junction and the thermal interface), on the area S_{JC}, and on the thermal conductivity k_{JC}. R_{JC} is given by the following:

$$R_{JC} = \frac{e_{JC}}{k_{JC} S_{JC}} \tag{6}$$

R_{CS}, the case to heatsink thermal resistance; a conductive resistance that takes into consideration the thickness and conductivity of the case and interface material. R_{CS} is given by the following:

$$R_{CS} = \frac{e_{CS}}{k_{CS} S_{CS}} \tag{7}$$

R_{SA}, *convection resistance* between the fins and air. R_{SA} is a convective resistance given by the following:

$$R_{SA} = \frac{1}{h S_F} \tag{8}$$

Where S_F is the fins surface area in contact with ambiance and h is the convection heat transfer coefficient between the fins and the ambient environment (see Appendix 1).

The sum $R_{JC} + R_{CS} + R_{SA}$ represents the *heat exchanger thermal resistance* between the junction and ambient. *The thermal resistance of the heat exchanger* is then given by the following:

$$R_{th} = R_{JC} + R_{CS} + R_{SA} = \frac{e_{JC}}{k_{JC} S_{JC}} + \frac{e_{CS}}{k_{CS} S_{CS}} + \frac{1}{h S_F} \tag{9}$$

Thus, the *heat exchanger thermal resistance* is a parameter, which depends on the materials constituting the heatsink, the casing, and the thermal interface material. It also depends on the surface area in contact with air, the configuration of the fins, their number, the position of the heatsink (horizontal/vertical), air temperature, etc. Its determination is somewhat complicated [6] and often necessitates the running of experiments.

Hopefully, when integrated systems are offered, values of the thermal resistance, R_{th}, are given by manufacturers' data. In the case where the heat exchanger is mounted from separate pieces, the heat exchanger thermal resistance can be computed from Eq. (9). Electronic component suppliers' data sheets give the parameters necessary for the computation of R_{CS}. Similarly, characteristics of gaskets

N°	Source	Heatsink	Thermal resistance R (°C/W)
1	[7]		R_{th} = 2.4°C/W
2	[8]		R_{th} = 17°C/W
3	[9]		R_{th} = 28°C/W
4	[10]		R_{th} = 55°C/W
5	[11]	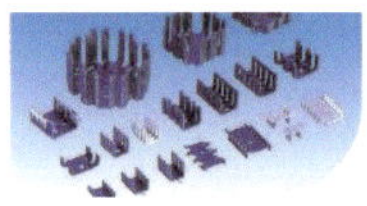	R_{th}: 7–80°C/W
6	[12]		R_{th}: 2–56°C/W
7	[13]		R_{th} = 11°C/W
8	[14]		R_{th} = 4.75°C/W
9	[15]		R_{th} = 63°C/W
10	[16]		R_{th} = 14°C/W

Table 1.
Values of R_{th} *for different heat sinks.*

(eventually), gap fillers, and other interface materials will permit the calculation of R_{CS}. Finally, R_{SA} can either be taken from heatsink data sheets (**Table 1** shows values of R_{SA} for different heatsink makes), or calculated from Eq. (8) and Appendix 1.

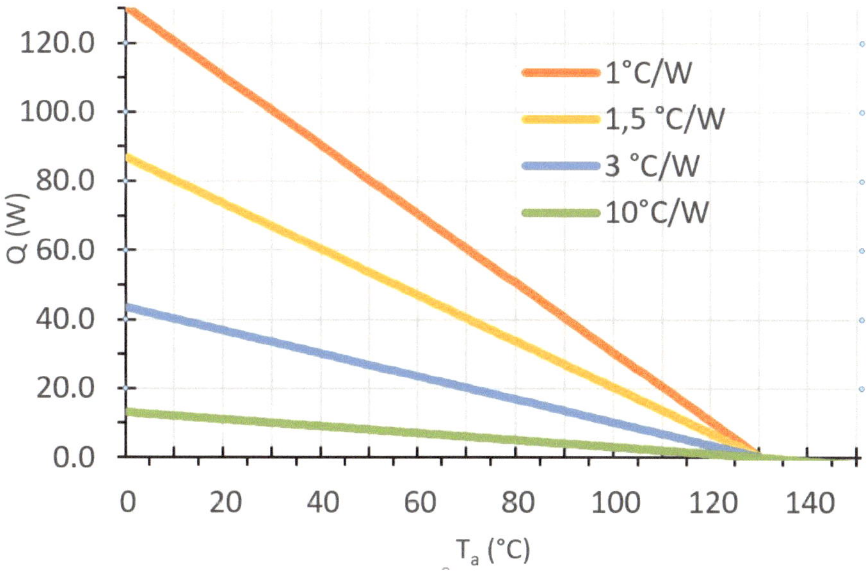

Figure 6.
Dissipation power as a function of ambient air temperature.

Remarks:

i. Sometimes, some manufacturers give two values of the thermal resistance of the heatsink. This can be the case in two situations:

- If the heatsink is such that it can be mounted vertically or horizontally, the lower value of thermal resistance corresponds to the *vertical mounting* of the fins, which does not get in the way of air movement from the low to the high parts of the fins. Heat transfer is therefore more efficient. In contrast by opposing air movement, horizontal assembly of the fins leads to a 20% of efficiency.

- If the gasket and the filling material resistances are not included, the second value corresponds to the resistance of these mounting accessories. In this case, one should add the two values to obtain the thermal resistance, $R_{CS} + R_{SA}$.

ii. In this chapter, R_{th}, refers to the total resistance of the *heat exchanger*, as shown in Eq. (9).

iii. It should be noticed that the smaller the *heat exchanger thermal resistance* is, the higher is the *dissipation power* Q, and the better is the heat exchanger and the cooling of the electronic equipment. **Figure 6** shows the evolution of the dissipation power as a function of the surroundings temperature for a component with T_{op} = 130°C, using different heat exchangers: R_{th} = 1°C/W to R_{th} = 10°C/W

3. Passive heat exchanger design

The simplest design technique uses the thermal resistance model of the heat exchanger. It supposes that the coolant (air for example) temperature, T_a, is constant in contact with the heatsink. Using this model, the power of heat dissipated by a heatsink is written as follows:

$$Q = \frac{T_J - T_a}{R_{th}} \tag{10}$$

Where:
Q is the heat flux from the component to ambient air (Watts),
T_J and T_a are, respectively, junction and air temperatures (°C),
and R_{th} is the thermal resistance of the heat exchanger (°C/W).
Three design problems can generally be encountered:

- *A sizing problem:* Given an electronic component or system with a maximum power generation P, how can one choose, among commercially available products, the heatsink that would ensure dissipation of the heat generated during operation?

- *A verification problem:* You have an electronic component/system with a heatsink mounted on. Is this heatsink correctly sized; that is, will it fit your electronic component or system?

- *Positioning circuit boards on a rack for optimal heat exchange:* How can a number of circuit boards are placed in a rack so that cooling is optimal?

3.1 Heat exchanger sizing

In this case, it is desired to find a commercially available heatsink, capable of dispersing the heat generated by the operation, in surroundings of temperature T_a, of an electronic component with a maximum power generation Q_{max} and a maximum junction operating temperature, T_{op}.

The sizing procedure is carried out in two steps.

STEP 1: From component, thermal interface, and heatsink datasheets select those which satisfy the space constraint. Get the values of Q_{max}, T_{op}, R_{JC}, R_{CS}, and R_{SA}.

STEP 2: Write Eq. (10) for Q_{max} and determine the thermal resistance of the heatsink which will insure operating temperature not exceed T_{op}.

Written for Q_{max} and T_{op} Eq. (10) gives the following:

$$Q_{max} = \frac{T_{op} - T_a}{R_{th}} \tag{11}$$

Rearranging, we obtain:

$$R_{th} = \frac{T_{op} - T_a}{Q_{max}} \tag{12}$$

Illustration 1

An electronic chip is required to operate at a surrounding temperature of 45°C. This chip is mounted with a casing having a thermal resistance R_{JC}, and with thermal interface materials having a global resistance R_{CS}. The data sheet of the chip shows that it can generate up to Q_{max} at T_{op}.

*Supposing there are no space or geometric constraints, what are the heat sinks that can be used among those of **Table 1**.*

Data:
$R_{JC} = 0.3°C/W$ $R_{CS} = 0.4°C/W$.
Q_{max} =10 W T_{op} = 85°C.
Solution
Using Eq. (12), the thermal resistance at maximum power is: $R_{th} = \frac{85-45}{10} = 4°C/W$
The adequate heatsink resistance is obtained using Eq. (9) as follows:

$$R_{SA} = R_{th} - \left(R_{JC} + R_{CS}\right)$$

$$R_{JC} + R_{CS} = 0.7\ ^{°C}\!/_{W} \Longrightarrow R_{SA} = 3.3\ ^{°C}\!/_{W}$$

This means that only heat sinks with thermal resistances lower or equal to 3.3°C/W will be able to give heat exchangers with $R_{th} \leq 4°C/W$*, thus ensuring safe operation of this chip.*

This is the case for references 1 and 6 only.

3.2 Heat exchanger verification

In this case, it is desired to verify if a given heatsink will permit the safe operation of an electronic component at a known surroundings temperature.

The verification procedure is carried out in four steps.

STEP 1: From data sheets (component, thermal interface, and heatsink) get the values of Q_{max}, T_{op}, R_{JC}, R_{CS}, and R_{SA}.

STEP 2: Calculate R_{th} from Eq. (12): R^*_{th}.

STEP 3: Calculate R^*_{SA} from Eq. (13) as follows:

$$R^*_{SA} = R^*_{th} - \left(R_{JC} + R_{CS}\right) \tag{13}$$

STEP 4: If $R_{SA} \leq R^*_{SA}$ then the heatsink will do the job.

Illustration 2

You bought an electronic component with a heatsink mounted on, thus constituting a heat exchanger ready to use. The heatsink is the one shown under N° 10 in ***Table 1***.

The data sheet of the electronic component gives the following information: Q_{max} = 10 W *and* T_{op} = 115°C.

Will this system operate safely at a surrounding temperature of 27°C?

Solution

STEP 1: Q_{max} = 10 W, T_{op} = 115°C *and*

N° 10 in ***Table 1*** *gives:* R_{th} = 14°C/W.

STEP 2: Eq. (11) gives: $R^*_{th} = \frac{115-27}{10} = 8,8°\frac{C}{W}$.

STEP 3: $R_{th} > R^*_{th}$*: the mounted heatsink will not do the job.*

3.3 PCBs optimal assembly

A typical passive heat exchange problem is encountered when one needs to assemble a set of circuit boards on a rack.

Figure 7 shows a number of boards of the same size, H, placed on a rack of length L and width w. The spacing between the different circuit boards is assumed to be constant, δ, whereas each board releases a constant flux density, q (W/m^2). The cooling of the circuit boards is ensured by natural convection.

In such a configuration, the temperature of the boards is not uniform, rather it increases with height, reaching its maximum value at H. The spacing δ needs to be such that heat evacuation is optimum.

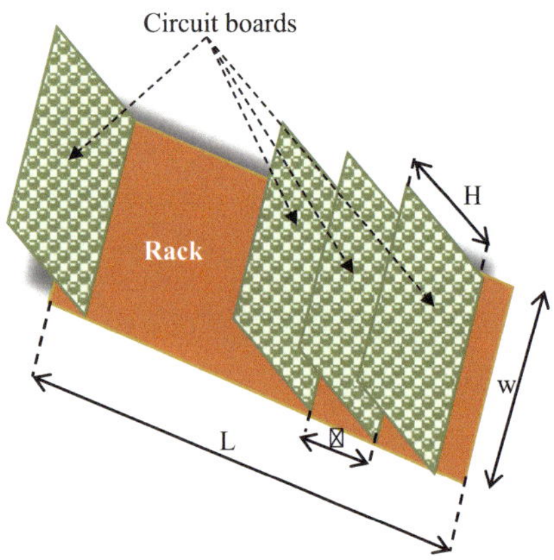

Figure 7.
Assembling a set of circuit boards on a rack.

3.3.1 Optimum spacing of PCBs

A frequent question is; *how should the boards be placed on the rack for optimum heat evacuation?* In other words, *what is the optimum spacing,* δ^**?*

The optimal spacing is given by [17]:

$$\delta^* = 2,12\frac{w}{Ra_q^{0,2}} \tag{14}$$

Ra_q being the Rayleigh number for, q, given by:

$$Ra_q = 2,12\frac{w^4 C_p \rho^2 g \beta q}{\mu k^2} \tag{15}$$

Where

- C_p, ρ, β, μ and k are respectively the specific heat, the density, the expansion factor, the viscosity, and the thermal conductivity of air[3].

- g gravity constant; $g = 9.81\ \mathrm{ms}^{-2}$

3.3.2 Optimal number of boards

The optimum number of boards of thickness e, to be assembled on a rack of length L, is then given by:

$$n^* = \frac{L}{\delta^* + e} \tag{16}$$

Other issues of importance in rack design such as the determination of heat transfer coefficients between electronic boards and air, and the calculation of the heat flux evacuated by natural convection are discussed in Ref. [1].

---

[3] See Data Bank in Ref. [18].

4. Active heatsink technologies

During the last decade, unprecedented technological developments imposed tremendous flows of information (data or multimedia), implying an *ever-growing need for fast data transfers* (Bluetooth, Wi-Fi) and for *larger computing and storage capacities*. In addition, the advent of 5G and the *Internet of Things* means that the number of connected devices is constantly increasing [19]. Moreover, *onboard electronics* and *mobile telephony* created new ergonomics and space (volume) constraints. From a systems engineering perspective, this implies that *larger power circuits have to be integrated in smaller volumes*.

Under such conditions, the amounts of energy generated by electronic circuits are so great that *no passive heatsink of a reasonable size will be able to do the cooling job*. This is because passive sinks dissipate heat mainly through natural convection, but natural convection is no longer sufficient to extract the heat generated by the operation of these power components. Thus, the use of active cooling systems, with or without refrigeration or phase change is necessary [20, 21] for the safe operation of power electronics.

Active cooling systems are heat exchangers where the flow of the heat transfer medium (air or liquid) is forced by a fan or a pump. Several technologies exist. They can be sorted into two categories:

i. single-phase heat sinks, and

ii. two-phase heat sinks.

4.1 Single phase

Single-phase heat sinks are heat exchangers in which the cooling fluid (liquid or gas) *does not undergo any phase change*, meaning that if the cooling fluid is a liquid, then it remains liquid throughout the cooling process.

Single-phase heat exchangers used for electronic systems cooling include the following:

- Forced convection heat sinks
- Cold plate heat sinks
- Microchannel coolers
- Refrigerated coolers

By far, forced convection heat sinks are the simplest.

4.1.1 Forced convection heat sinks

Forced convection heat sinks are typically heat exchangers formed of finned surfaces similar to those presented in **Table 1**, where cooling air is forced through the fins using *fans*.

Figure 8 shows such a heat exchanger fixed on a CPU board. The fan is mounted on top of the heatsink to draw air through the fin surfaces.

Depending on the type of fins configuration, the fan can also be mounted in the middle of the finned heatsink as shown in **Figure 9**.

Figure 8.
A fan on a heatsink mounted on a CPU board. Source [22].

Figure 9.
Fan in the middle of a finned heatsink. Source [23].

Figure 10.
Miniaturized forced convection heatsink module. Source [24].

Very often, the assembly fins-fan is miniaturized in a cooling module similar to the one presented in **Figure 10**.

Forced convection heat exchangers become necessary when Q_{max} exceeds 70 W/cm^2 [25]: Using a fan to force air through the finned heatsink increases significantly heat transfer to air; *convective heat transfer coefficients* can reach values as high as 3000 $Wm^{-2o}C^{-1}$ [26]. For comparison, this coefficient is between 25 and 500 $Wm^{-2o}C^{-1}$ for natural convection heat sinks [1].

We should underline, however, that including a fan in an electronic system will require an additional power supply. This will certainly impact the final size of the system.

4.1.2 Forced convection heatsink design

An important parameter in the design of a forced convection heatsink is the flow rate of the cooling fluid (e.g., air) going through the fins of the sink [2]. As a matter of fact, the flow rate generated by the fan actually impacts the heat transfer coefficient: A higher flow rate will mean a greater cooling fluid velocity, leading to a better convection heat transfer [1]. The flow rate also impacts the pressure drop across the heatsink given that for a given fin configuration, higher flow rates generate higher pressure drops[4].

Consequently, the proposed design procedure will search for *a balance between a reduced thermal resistance and an acceptable pressure drop*. It can be organized in the following seven-step procedure.

STEP 1: Collect heatsink pressure drop data.

Flow over a fanned heatsink generates a pressure drop that depends on several parameters including the number of fins, the distance between two fins, the value of the flow rate, etc. [26]. Suppliers of heat sinks will generally be able to provide, for the different models they propose, plots of the pressure drop versus the cooling fluid (air) flow rate. These plots are referred to as *heatsink working diagrams* where the flow rate is generally expressed in cubic feet per minute (CFM: ft^3/mn).

Figure 11 shows an example of such plots.

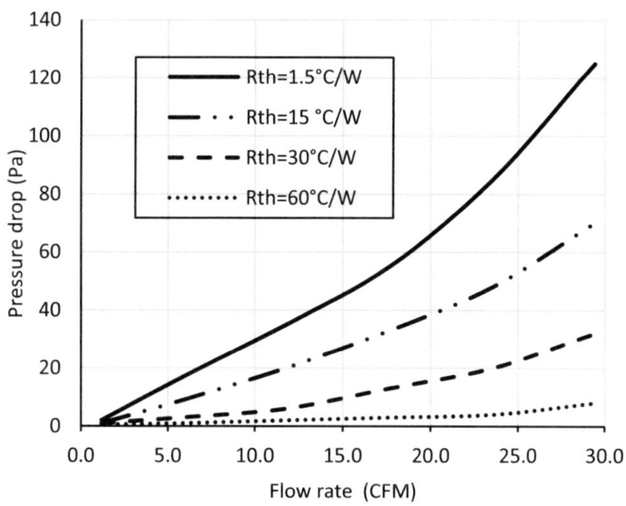

Figure 11.
Heatsink working diagram.

[4] For laminar flow, the pressure drop is proportional to the square of the flow velocity [26].

STEP 2: Get the fan curves.

Similarly, fan suppliers provide plots that give, for each model proposed, the flow rate achievable by the fan under a given pressure drop. These plots are often referred to as *performance curves of the fans*.

Figure 12 presents examples of performance curves for four fans.

STEP 3: Select heatsink candidates based on their working diagrams.

The selection criterion here is the individual thermal resistance, R_{th}^{s}.

STEP 4: Define the allowable coolant temperature rise.

The *coolant temperature rise* is defined as follows:

$$\Delta T = T_o - T_i \tag{17}$$

where T_i and T_o are, respectively, the inlet and outlet temperatures of the coolant.

T_i being known (generally, the coolant enters the heatsink around ambient temperature: $T_i \approx T_a$), the definition of ΔT permits to calculateT_o:

$$T_o = T_i + \Delta T \tag{18}$$

STEP 5: Determine coolant flow rate for each heatsink candidate.

For each heatsink s under evaluation, the thermal resistance, R_{th}^{s}, is known. This makes it possible to calculate the coolant flow rate by combining Eq. (9) and a heat balance on the cooling fluid:

Equation (9) gives:

$$Q_s = \frac{T_c - T_a}{R_{th}^{s}} \tag{19}$$

The heat balance for the cooling fluid gives the following:

$$Q_s = \dot{m}_s C_p (T_o - T_i) = F_s \rho C_p \Delta T \tag{20}$$

Where:

- $\dot{m}_s$ is the mass flow rate of the coolant fluid (air) over the heatsink s, kg/sec,
- C_p is its sensible heat (J/kg°C),

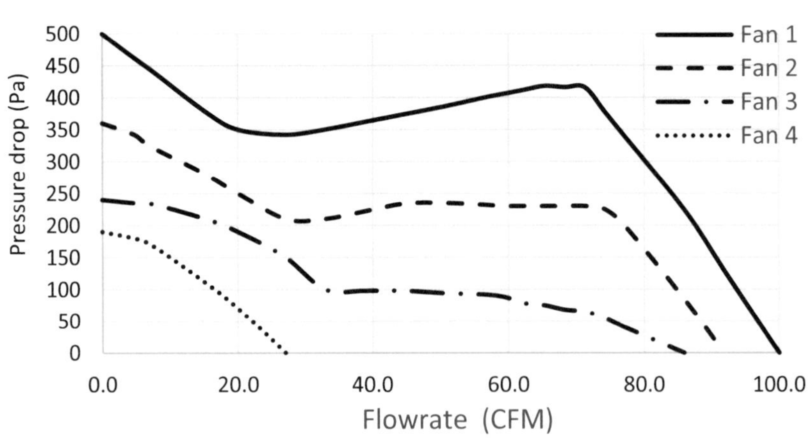

Figure 12.
Performance curves of fans.

- ρ is its density (kg/m^3), and
- F_s is the coolant volume flow rate (m^3/s).

Substituting for Q_s in Eq. (19) and extracting the coolant volume flow rate, F_s:

$$F_s = \frac{T_c - T_a}{\rho C_p \, \Delta T \, R_{th}^s} \tag{21}$$

STEP 6: Determine the pressure drop for each of the heatsink candidates.

Injecting the values of flow rates into the working diagram gives the pressure drop which will be generated by each of the heatsink candidates. **Figure 13** shows this procedure for the first two heat sinks considered in **Figure 10** (R_{th} = 1.5°C/W and R_{th} = 15°C/W), which permits the determination of pressure drops ΔP_1 and ΔP_2 generated by when F_1 and F_2 flow over sink 1.

STEP 7: Determine compatible fans using the performance curves.

For each heatsink candidate, s, inject the pressure drop, ΔP_s, determined in the previous step in the performance curves. This will generate the series of flows, $F_s^{fan_i}$, which will be delivered by fan i, operating against the pressure drop ΔP_s.

Figure 14 shows that, for fan 1, all performance curves cross the line ΔP_1. This means that any of the fans under evaluation can be used. The choice will then be made on price and volume criteria.

4.1.3 Cold plate heat exchangers

Cold plate heat sinks could be considered as a particular class of *plate heat exchangers* [27]. They are used when thermal powers released by electronic systems (smartphones, electric cars, onboard avionics systems, TGV, etc.) become so important that forced convection heat sinks are no longer sufficient [28–30]. They are constituted of plates that are fitted with pipes (see **Figure 15**) through which cooling fluid passes. They are attached to the surfaces of the electronic component or to the board to be cooled. The coolant is conveyed by means of a pump. The coolant itself is, in turn, cooled using a compact exchanger [27].

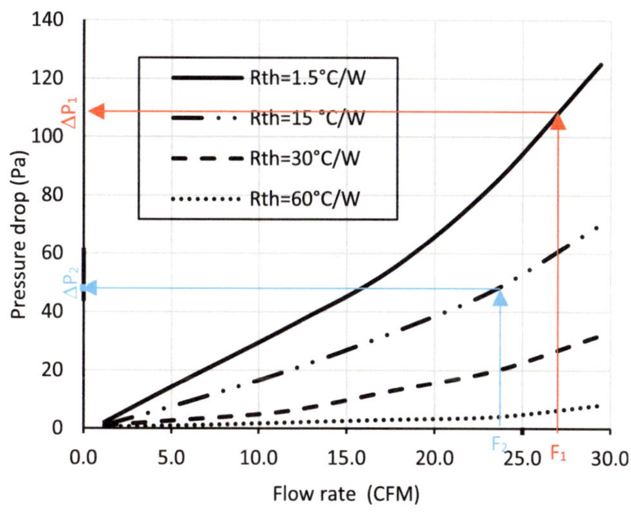

Figure 13.
Determination of pressure drops for heat sinks 1 and 2.

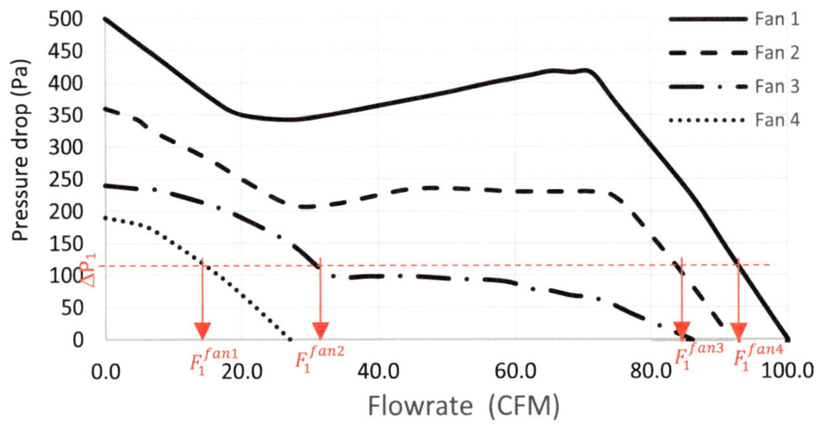

Figure 14.
Determination of deliverable flows under ΔP_1.

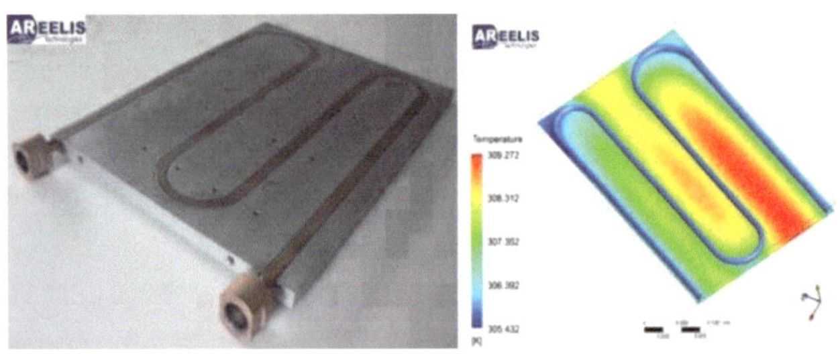

Figure 15.
A cold plate heatsink and its thermal scan. Sources [28, 29].

4.1.4 Microchannel heat exchangers

Microchannel heat sinks can be considered as a particular subclass of *printed circuit heat exchangers* [27]. These are indeed very small exchangers with overall dimensions not exceeding a few millimeters.

In contrast to their very small dimensions, they allow heat transfers in the order of 800 W/cm^2 [30–33]. The cooling fluid (generally air) circulates in microchannels, of microscopic equivalent diameters (approximately 10–60 μm), formed by etching on metal plates or in composite materials [34]. These microchannels have heights of the order of 0.5 mm. The modules thus formed are placed under the electronic components to be cooled [35]. **Figure 16** illustrates such an assembly where the circuit board to be cooled is shown in semi-transparency, above the microchannels.

However, microchannel cooling suffers from several drawbacks: complex implementation, significant pressure drops associated with microchannel flow. Moreover, it does not quite meet the requirements of thermal management in power electronic systems [36].

4.1.5 Refrigerated heat exchangers

Significant efforts have been devoted during the last 10 years to the development of a new kind of heat sinks. These are constituted of heat exchange plates associated with refrigeration cycles and a refrigerant as the heat transfer fluid [37–39]. Such refrigerated heat exchangers are used to cool power electronic

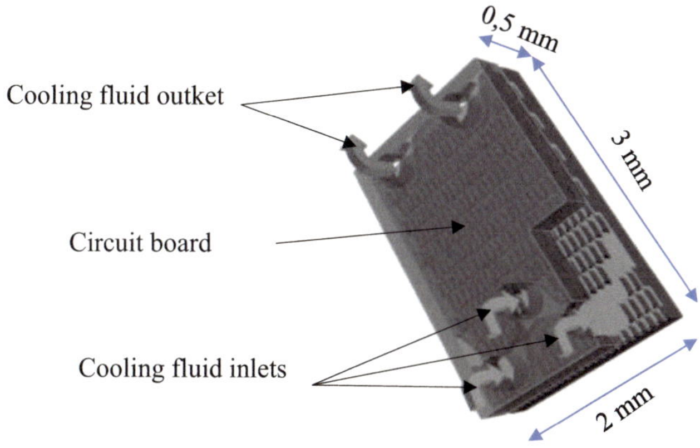

Figure 16.
Overall dimensions of a micro-channel heatsink.

systems such as laser power supplies and their optical systems. They are able to extract heat flux densities exceeding 1000 W/cm^2 while keeping microchips at temperatures below 65°C [39–44].

Note, however, that mounting this type of heat exchanger on electronic equipment remains difficult due to the fact that all the components of a refrigeration cycle (compressor, expansion valve, evaporator, and condenser) must be assembled in rather tiny spaces (**Figure 17**).

These systems must therefore meet an important challenge: the miniaturization required by designs favoring small sizes [21, 42, 43].

4.2 Two-phase heat sinks

These are phase change micro-exchangers, where the cooling fluid (generally a liquid) undergoes phase changes (from liquid to vapor and back to liquid) during the heat transfer process. Two-phase heat sinks are generally constituted of a vessel containing a heat transfer fluid; water in most cases. The lower surface of the vessel is in contact with the heat source (electronic equipment), and the upper surface is in contact with the cooling medium (generally air). **Figure 18** presents the operating principle of this system.

When in contact with the lower surface (heat source), the heat transfer fluid evaporates by absorbing the heat generated at this surface. The vapor then goes

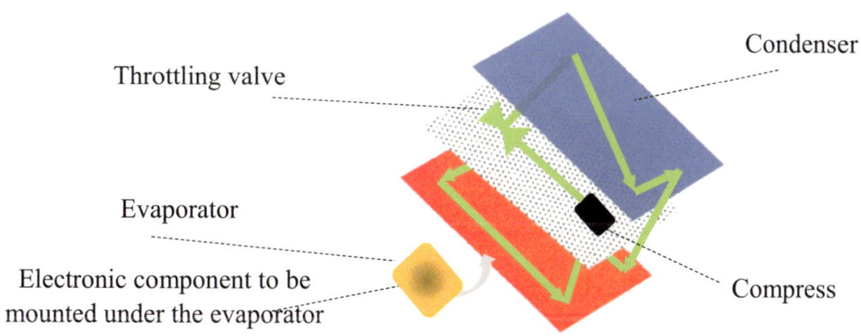

Figure 17.
Exploded view of a chip mounted under a refrigerated heatsink.

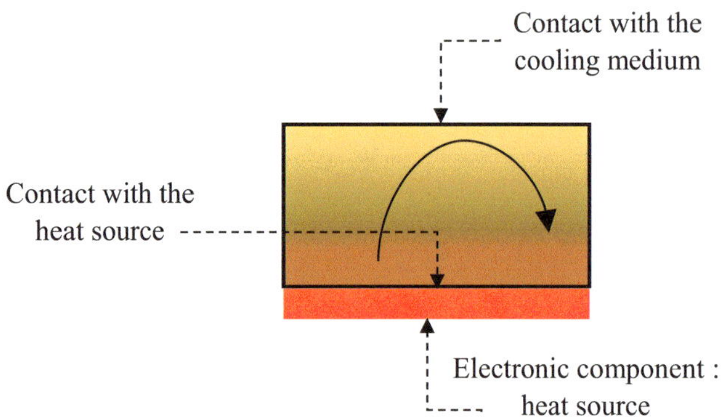

Figure 18.
Operating principle of a two-phase heatsink.

upwards, toward the upper surface where it gets cooled and condensates. Condensed liquid falls back along the walls of the vessel, down to the lower surface, where it absorbs again the heat and evaporates. This cycle is repeated as long as heat is generated by the electronic equipment.

This way, two-phase heat sinks are used to remove heat from the heat source to the ambient environment. The lower surface acts as an evaporator and the upper one plays the role of a condenser.

Note that two-phase heat exchangers involve latent heat instead of sensible heat in energy transfers, thus allowing large amounts of heat to be exchanged over small areas, which leads to great compactness.

Two-phase heat sinks include the following:

- Spray coolers
- Vapor chambers
- Heat pipes.

4.2.1 Spray coolers

Spray coolers are a special class of two-phase heat exchangers since the heat transfer fluid undergoes a series of evaporations and condensations [45, 46]. The electronic component is cooled by a jet of fluid, which partially evaporates while absorbing heat (see **Figure 19**). The heat transfer fluid is sprayed directly onto the surface of the power component to be cooled [47–49].

Spraying enhances the vaporization of the fluid even at relatively low temperatures (60–75°C, under 1 Atm.). It is generally realized using a single injector [50] or multiple injectors [51].

However, like refrigerated exchangers, the realization of spray cooling is relatively complex. As shown in **Figure 20**, it requires, in addition to the sprayers, the installation of a condenser to collect the vapors generated and a pump or compressor to feed the injectors.

4.2.2 Vapor chambers

Vapor chambers are the most common two-phase-heat exchangers where a heat transfer fluid (water in most cases) is placed in a sealed envelope: *the chamber* [52]. The lower surface of the chamber is mounted on the electronic component to be

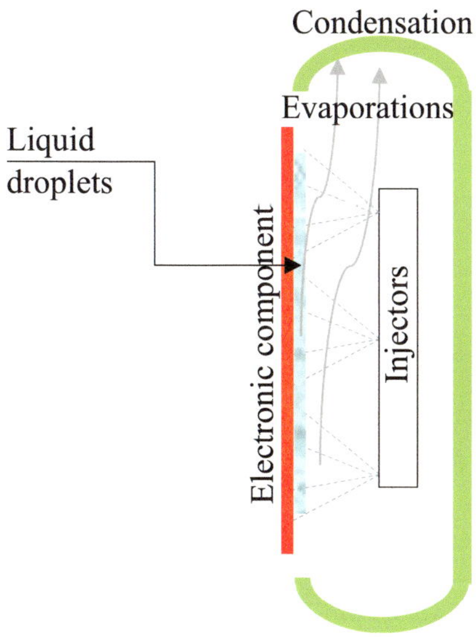

Figure 19.
Spray cooling system.

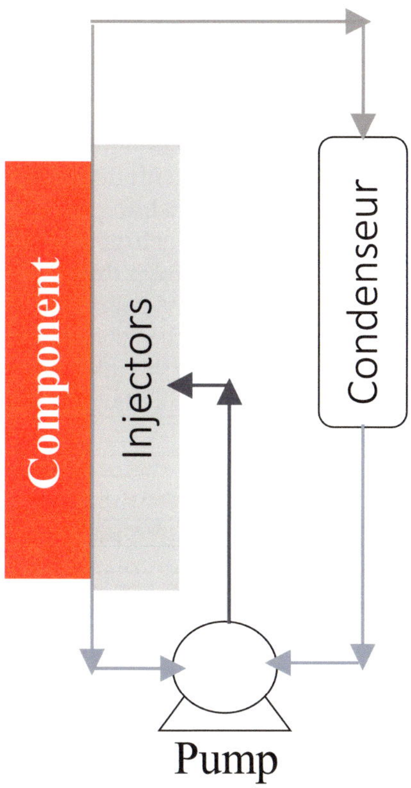

Figure 20.
Spray cooling cycle.

cooled. As described above, the liquid absorbs the heat generated by this component to evaporate. The vapors then condense on the cold surface of the chamber, which transfers the ambient environment the energy liberated by the condensation process [53, 54].

Figure 21.
Vapor chamber heat exchanger. Source [55].

Figure 21 shows this type of heat exchanger. It should be noted that the dimensions are extremely small to meet the requirements of miniaturization of onboard electronics and mobile telephony. The typical thickness of such an exchanger is 2–8 mm.

Current developments focus on the ultra-miniaturization of these exchangers by introducing ultra-thin evaporation chambers: thicknesses from 0.3 to 2 mm using walls made of titanium, stainless steel, or copper alloys. **Table 2** shows the current uses of these types of exchangers, as well as their typical dimensions and the thermal powers conveyed.

4.2.3 Heat pipes

A heat pipe consists of a sealed tube, containing a heat transfer fluid, without any other gas (see **Figure 22**). In most cases, only a small quantity of water suffices: About 1 cc of water for a 150 mm long, 6-mm heat pipe is typical.

In one of its zones (vaporization zone), the heat pipe tube is in contact with the hot source to be cooled. The heat recovered from this source increases the temperature of the heat transfer fluid causing it to evaporate.

The resulting vapors then accumulate in the condensation zone of the heat pipe where they condense on the internal walls of the tube, releasing their latent heat to the ambient environment (see **Figure 22**). The condensate flows in droplets on

Domain of use	Thickness (mm)	Length (mm)	Width (mm)	Power range (W)	Construction materials
Smart phones	0.3–0.4	80	50	5	Copper alloy or stainless steel
Portable microcomputers	0.5	220	60	15–25	Copper alloy
Base telecommunication stations	2–4	100	100	50–100	Copper or copper alloy
Graphic boards	2–4	200	90	200–350	Copper or titanium
Servers	3–4	150	80	200–400	Copper
Game consoles	3–5	150	150	150–200	Copper
Power inverters	3–8	450	450	500–2000	Copper

Table 2.
Vapor chamber uses and construction materials.

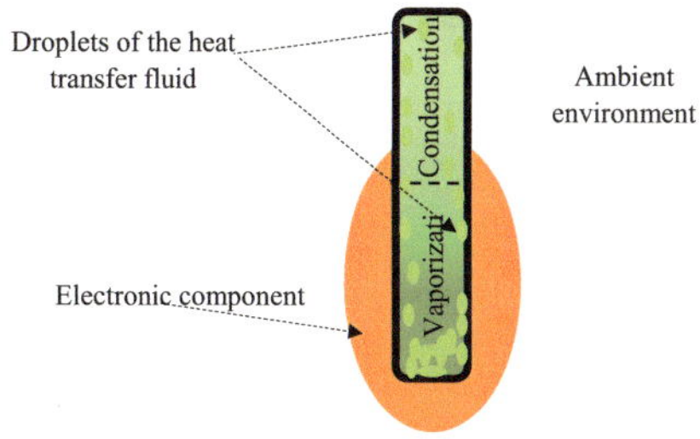

Figure 22.
Heat pipe working principle.

the walls of the tube and returns to the vaporization zone to be again submitted to the heat flow and to evaporate.

4.2.4 Heat pipe exchanger

These special heat exchangers are generally formed of a number of heat pipes usually made of copper, where several fins (the fin stack) have been mounted in the condensation zone.

As shown in **Figure 23**, each pipe is bent in a U form, slightly flattened where contact is made with the mounting base plate (hot area), and filled with a heat transfer fluid. When the copper pipes are going to be directly in contact with the environment, protection against corrosion could be obtained by nickel plating.

Each heat pipe acts actually as a *heat mover* from the area where heat is generated (the base plate) to the cooling area (fin stack) where air movement carries the heat away.

Rules of thumb

1. To maximize the amount of energy received from the heat source (the electronic component), each heat pipe should be placed directly above this source. The closer and tighter the contact is, the better is the heat transfer.

2. To improve the contact between the source and the heat pipe, sections of the pipes that are in contact with the source are flattened to about 30–65% of their original diameter.

Figure 23.
Fixing the fin stack to the base plate.

3. The bend radius of the heat pipes is to be fixed with care in order not to alter the integrity of the pipes.

4. The minimum *bend radius* should be *three times the diameter of the heat pipe*.

Note that the heat pipes can be directly fixed on the electronic component to be cooled. **Figure 24** shows a heat exchanger made of six copper heat pipes that are directly fixed on a central processing unit (CPU) to convey heat to a 13 fin stacks. The advantage is a direct contact between the component and the heat pipes, thus reducing the thermal resistance of the base plate.

A more elaborated version is the U-shaped vapor chamber design where a U-shaped heat plate replaces the heat pipes, the heat plate operating in the same way as flattened pipes. **Figure 25** shows a heat exchanger formed from a vapor chamber bent to offer a large copper base to be in direct contact with the power electronic component. This design offers the advantage of a large direct contact area, which translates into a better performance: more than 20% better than the design of the pipe.

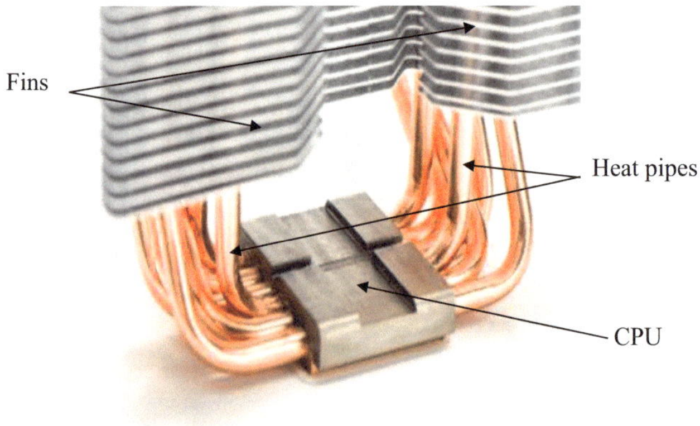

Figure 24.
Heat pipe heatsink to cool a CPU.

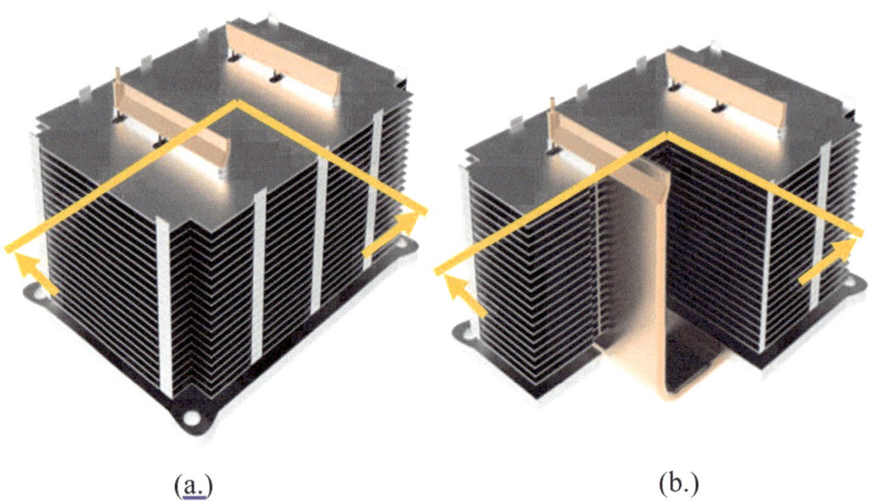

Figure 25.
U bent vapor chamber with a fin stack. (a) Complete vapor chamber and (b) showing inside the chamber.

4.3 Two-phase heat exchanger design

4.3.1 Design fundamentals

4.3.1.1 Carrying capacity

The performance of a two-phase heatsink is measured by its carrying capacity, Q, defined below.

Definition

The *carrying capacity*, Q, of a two-phase heat exchanger is defined as the amount of heat the device can move out per unit time.

From its definition, Q is an *energy per unit time*, which is a *power* which should be expressed in *Watts*.

For a given two-phase heat exchanger, the value of Q depends on several parameters: the size of the vessel, the latent heat of the transfer liquid used, the thermal conductivity of the metal constituting the vapor chamber or the heat pipe, convections (inside and outside the vessel), etc. However, we should note that, in two-phase heat sinks, energy is essentially transferred through the vaporization-condensation process, involving latent heat.

4.3.1.2 Two-phase thermal budget

Definition

The *thermal budget*, ΔT, of a two-phase heat exchanger is defined as the difference between the temperature of the electronic component (heat source), T_c, and the temperature of the ambiance where the component will operate, T_a:

$$\Delta T = T_c - T_a \tag{22}$$

The carrying capacity, Q, is proportional to the thermal budget ΔT (see Eq. (9)).

Thus for low thermal budgets, it will be difficult to achieve high values of Q. In this case, two-phase heat sinks should be used, mainly when $\Delta T < 40°C$.

4.3.1.3 Effective thermal conductivity

Heat exchange in two-phase sinks is actually complicated to model. It involves several heat transfer mechanisms: conduction through the vapor chamber or the heat pipe metal envelope, evaporation, condensation, convection inside and outside the vessel, etc. The exact representations of these transfers are interesting on the theoretical level of analysis. But, for sizing and design purposes, a more practical approach has been adopted by manufacturers and system designers. This approach is based on the thermal resistance model, in a way similar to that presented in Subsection 2.1 for single-phase heat sinks.

Let us recall that in conduction and in convection, thermal resistance models permit to express the power transferred Q in terms of R_{th} and the thermal budget, ΔT, as follows [2, 47]:

$Q = \frac{\Delta T}{R_{th}}$, where, R_{th} is the thermal resistance, which is expressed differently for conduction and convection:

- For a convection heat transfer:

$$R_{th}^{conv} = \frac{1}{hA} \tag{23}$$

Where h is the convective heat transfer coefficient and A the heat transfer area.

- For a conduction heat transfer:

$$R_{th}^{cond} = \frac{L}{kA} \tag{24}$$

Where L is the material thickness, A the heat transfer area, and k is the *thermal conductivity*.

Similarly, two-phase heat exchanger energy transfers are represented using an *effective thermal resistance*, R_{eff}, such that the carrying capacity, Q, is given by:

$$Q = \frac{\Delta T}{R_{eff}} \tag{25}$$

With $R_{eff} = \frac{L}{A\ k_{eff}}$, where k_{eff} is the *effective thermal conductivity* of the two-phase heat exchanger and A is its *heat transfer area of the heat sink*.

Definition

The *effective thermal conductivity*, k_{eff}, of a two-phase heat exchanger is defined as follows:

$$k_{eff} = \frac{QL}{A\Delta T} \tag{26}$$

Where Q is the *carrying capacity*, A is its *heat transfer area*, ΔT is the thermal budget (temperature difference between evaporator and condenser sections) and L is the *effective length* of the vessel which is the *distance from the midpoint of the evaporator to the midpoint of the condenser*.

It should be noted that, unlike the thermal conductivity of materials, the *two-phase heat exchanger thermal conductivity* varies with length. It is not a physical property of the material, but it is a modeling representation of heat transfer in two-phase heat sinks.

This modeling—using effective thermal conductivity, k_{eff}—makes it possible to represent *two-phase heatsink energy transfers* by *a simple thermal resistance model*, just the same way thermal resistance was used in single-phase heat sinks.

Effective thermal conductivities of two-phase heat exchangers range from 1500 to 50,000 W/m°C. For comparison, the thermal conductivities of the best energy conductors are around 400 W/m°C. Effective thermal conductivities of two-phase heat exchangers can reach 10–100 times the conductivity of the best thermal conductors like copper.

4.3.2 Design steps

Generally, manufacturers of two-phase heat exchangers supply, among other information on their product, the following parameters:

- Operating temperature limits: generally between 20 and 250°C.

- The heat exchanger carrying capacity.

- The heat exchanger thermal resistance.

The following five-step procedure is proposed to design two-phase heat exchangers.

STEP 1: The datasheet of the electronic component to be cooled gives the power and operating temperature: Q (W) and T_{op} (°C).

STEP 2: Knowing ambient temperature, T_a, calculate the thermal budget ΔT as follows:

$$\Delta T = T_{op} - T_a \tag{27}$$

STEP 3: Calculate the *design thermal budget* ΔT_D as follows:

$$\Delta T_D = \Delta T - 5 \tag{28}$$

STEP 4: Divide the *design thermal budget*, ΔT_D, by the power, Q, of the electronic equipment to be cooled. This division will give the maximal thermal resistance to be assured by the two-phase heatsink: R_{Max}

STEP 5: Select the two-phase heatsink having a thermal resistance $R < R_{Max}$.

Rules of thumb

1. Derate the thermal budget by 5°.

2. In the case of heat pipes, if bending is considered, derate the carrying power by 2.5% for every 45-degree bend.

3. When heat pipes are flattened to ensure a better contact with the base plate, the flattening should be in the range of 30–65%.

4. Flattening implies a 15–30% reduction in the carrying capacity of the heatsink.

A. Appendix 1

Determination of heat transfer coefficients for electronic components and circuit boards under natural convection

A.1 Electronic components under natural convection in air

$$\mathrm{h} = 3.53\left(\frac{\mathrm{T_w} - \mathrm{T_a}}{\mathrm{H}}\right)^{0.25} \tag{29}$$

Where:

H is the height of the small electronic component, *expressed in meters*

T_w and T_a are, respectively, the wall temperature and the ambient temperature

A.2 Circuit boards under natural convection in air

$$\mathrm{h} = 2.44\left(\frac{\mathrm{T_w} - \mathrm{T_a}}{\mathrm{H}}\right)^{0.25} \tag{30}$$

Where:

H is the height of the electronic board considered, *expressed in meters*

T_w and T_a are, respectively, the board and the ambient temperatures.

Author details

Abdelhanine Benallou
SIGMA TECH, Rabat, Morocco

*Address all correspondence to: pr.benallou@gmail.com

References

[1] Benallou A. Energy Transfers by Convection (Energy Engineering Set—Volume 3). NJ, USA: John Wiley & Sons Inc.; 2019

[2] Wilcoxon R, Martin G. Fan cooling—Part 1: Determining flow rate. In: Electronics Cooling Magazine. PA, USA: Lectrix, Spring; 2021. pp. 6-9. Available from: https://www.electronics-cooling.com/wp-content/uploads/2021/03/Electronics-Cooling_Spring-2021.pdf

[3] https://encryptedtbn0.gstatic.com/images?q=tbn:ANd9Gc RQ6FHjqIqAQ3svtm02NWba vJzOMuFOyviRQ&usqp=CAU

[4] https://ae01.alicdn.com/kf/HTB1e 3UmoY_I8KJjy1Xaq6zsxpXaK/Round-D133mm-Pre-drilled-led-pin-fin-heatsink-fit-for-CREE-cob-cxb3590-Bridgelux-V29-50.jpg_Q90.jpg_webp

[5] Nuts and Volts: https://www.nutsvolts.com/uploads/wygwam/NV_0709_Post_Fig 2_1.jpg

[6] Simons RE. Estimating parallel plate-fin heat sink thermal resistance. Electronics Cooling. 2003;**9**(1):8-9

[7] Sonelec: http://www.sonelec-musique.com/images/radiateur_to3_ml25_simple.jpg

[8] Sonelec: http://www.sonelec-musique.com/images/radiateur_to220_ml24.jpg

[9] Sonelec: http://www.sonelec-musique.com/images/radiateur_to5_co180.jpg

[10] Sonelec: http://www.sonelec-musique.com/images/radiateur_to5_ml61.jpg

[11] Mikkelsen-electronics: http://www.mikkelsen-electronics.com/images/FingerKK.jpg

[12] Myheat sinks: https://myheat sinks.com/docs/images/standard-heat-sinks/cold-forged/round-pin-heat-sink.jpg

[13] https://asset.conrad.com/media10/isa/160267/c1/-/en/183947_BB_00_FB/image.jpg?x=600&y=600

[14] Fischerelektronik: https://www.fischerelektronik.de/fileadmin/_migrated/pics/aufsteckgr.jpg

[15] https://cdn.webshopapp.com/shops/297304/files/322145950/800x1024x2/to-5-heatsink-63c-w-9mm-x-127mm-skk56-fisher.jpg

[16] https://i.ebayimg.com/images/g/3R4AAOSwTKZfMltz/s-l640.jpg

[17] Bar-Cohen A, Rohsenow W. Thermally optimum spacing of vertical natural convection cooled parallel plates. Journal of Heat Transfer. 1984; **106**:116

[18] Benallou A. Mass Transfers and Physical Data Estimation (Energy Engineering Set—Volume 1). NJ, USA: John Wiley & Sons Inc.; 2019. pp. 179-180

[19] Amalfi RL, Enright R, Kafantaris V. Emerging 5G Networks: Potential Economic Benefits of Two-phase Thermal Management. Electronics Cooling Magazine. PA, USA: Lectrix; June 2021. pp. 19-23. Available from: https://www.electronics-cooling.com/wp-content/uploads/2021/07/Electronics-Cooling-Summer-2021.pdf

[20] Zhang H, Che F, Lin T, Zhao W. Modeling, Analysis, Design, and Tests for Electronics Packaging Beyond Moore. Woodhead Publishing, UK: Elsevier Science; 2019

[21] Phelan PE, Chiriac VA, Lee T-YT. Current and future miniature refrigeration cooling technologies for high power microelectronics. IEEE Transactions on Components and Packaging Technologies. 2002;**25**: 356-365

[22] https://upload.wikimedia.org/wikipedia/commons/thumb/2/25/AMD_heatsink_and_fan.jpg/220px-AMD_heatsink_and_fan.jpg

[23] https://2.imimg.com/data2/WI/RM/MY-615568/pin-type-heat-sink 500x500.jpg

[24] Advanced Thermal Solutions Inc.: https://www.mouser.fr/images/marketingid/2019/microsites/122375246/Quad%20Flow%20Dual%20Flow.png

[25] PRIATHERM. Design your cooling system using forced convection in just three easy steps. Available from: https://priatherm.com/design-your-cooling-system-using-forced-convection-in-just-three-easy-steps/ [Accessed: 13 December 2018]

[26] Simons RE. Estimating parallel plate-fin heat sink pressure drop. In: Electronics Cooling Magazine. PA, USA: Lectrix; April 2016. Available from: h ttps://www.electronics-cooling.com/2016/04/calculation-corner-estimating-parallel-plate-fin-heat-sink-pressure-d rop/ [Accessed: 24 July 2021]

[27] Benallou A. Série Ingénierie de l'Énergie, volume 6, Échangeurs de Chaleur: Conception et Algorithmes de Calcul. Section 1.5, ISTE Editions. London; 2021. pp. 5-16 (under press)

[28] AREELIS Technologies: https://www.areelis.fr/wpcontent/uploads/2019/05/plaques_froides-495x400.jpg

[29] AREELIS Technologies: https://www.areelis.fr/wp-content/uploads/2019/05/FRE-0814-01-2-495x400.png

[30] Milnes P.D. Phase separation in two phase microfluidic heat exchangers [PhD dissertation]. Stanford University; March 2011 Dissertation available online at: http://purl.stanford.edu/ns144db9649

[31] Roland Baviere. Etude de l'Hydrodynamique et des Transferts de Chaleur dans des Microcanaux. Dynamique des Fluides [physics.flu-dyn]. PhD dissertation Diffusée par Université Joseph-Fourier - Grenoble I, 2005

[32] Meysenc L. Étude des micro-échangeurs intégrés pour le refroidissement des semi-conducteurs de puissance [PhD dissertation]; France: Institut National Polytechnique Grenoble; 1998. pp 117-123

[33] Tuckerman DB, Pease RFW. High performance heat sink for very-large-scale integrated circuits. IEEE Electron Device Letters. 1981;**2**(5):126-129

[34] Ohadi M, Choo K, Dessiatoun S, Cetegen E. Next Generation Microchannel Heat Exchangers. Nature Switzerland: Springer; 2013

[35] Colgan EG, Furman B, Gaynes M, Graham WS, LaBianca NC, Magerlein JH, et al. A practical implementation of silicon microchannel coolers for high power chips. IEEE Transactions on Components and Packaging Technologies. 2007;**30**(2):218-225

[36] ITRS. International Technology Roadmap for Semiconductors, 2007 Edition. Washington, DC: International Roadmap Committee. Semiconductor Industry Association; 2007

[37] Marcinichen JB, Lamaison N, Thome JR. Electronic micro-evaporator cooling systems and flow control. In: Thome JR, Kim R, editors. Encyclopedia of Two-Phase Heat Transfer and Flow. Singapore: World Scientific Publishing; 2016

[38] Webb RL. Next generation devices for electronic cooling with heat rejection to air. Journal of Heat Transfer. 2005;**127**(1):2-10

[39] Burnett J. Advances in vapor compression electronics cooling.

Electronics Cooling Magazine. PA, USA: Lectrix; June 2014. pp. 28-32. Available from: http://aspensystems.com/wp-content/uploads/2014/07/Electronics-Cooling-June-2014-Aspen-Article.pdf

[40] Barbosa JR Jr, Ribeiro GB, de Oliveira PA. A state-of-the-art review of compact vapor compression refrigeration systems and their applications. Heat Transfer Engineering. 2012;**33**(4–5):356-374

[41] Phelan PE, Catano J, Michna G, Gupta Y, Tyagi H, Zhou R, et al. Energy efficiency of refrigeration systems for high-heat-flux microelectronics. Journal of Thermal Science and Engineering Applications. 2010;**2**(3):1-6

[42] Shannon MA, Phillpott ML, Miller NR, Bullard CW, Beebe DJ, Jacobi AM, Hrnjak PS, Saif T, Aluru N, Sehitoglu H, Rockett A, Economy J. Integrated mesoscopic cooler circuits (IMCCS). In: Proc. ASME-AES. Vol. 39. Nashville, TN: International Mechanical Engineering Congress and Exhibition; 1999. pp. 75-82

[43] de Oliveira P A., Barbosa J R. Jr., Thermal Design of a Spray-Based Heat Sink Integrated with a Compact Vapor Compression Cooling System for Removal of High Heat Fluxes. Heat Transfer Engineering, 2015;36:14-15, 1203-1217

[44] Mancin S, Zilio C, Righetti G, Rossetto L. Mini vapor cycle system for high density electronic cooling applications. International Journal of Refrigeration. 2013;**36**(4): 1191-1202

[45] Kim J. Spray cooling heat transfer: The state of the art. International Journal of Heat and Fluid Flow. 2007; **28**(4):753-767

[46] Cheng WL, Zhang WW, Chen H, Hu L. Spray cooling and flash evaporation cooling: The current development and application. Renewable and Sustainable Energy Reviews. 2016;**55**:614-628

[47] Shicheng J, Yibin J, Liping P, Yu Z, Wenxuan Y. Investigation of spray cooling performance on different heat transfer surfaces. In: Proceedings of the International Conference on Logistics, Engineering, Management and Computer Science. Atlantis Press; 2014

[48] Thibault D. Étude du refroidissement par impact de jets à travers une paroi mince et avec un écoulement cisaillant amont: application aux aubes de turbines [PhD dissertation]. École Nationale Supérieure de Mécanique et d’Aérotechnique. Poitiers, France; 2009

[49] Kercher D, Lee JB, Brand O, Allen M, Glezer A. Microjet cooling devices for thermal management of electronics. IEEE Transactions on Components and Packaging Technologies. 2006;**26**:359-366

[50] Horacek B, Kiger K, Kim J. Single nozzle spray cooling heat transfer mechanisms. International Journal of Heat and Mass Transfer. 2005;**48**(8): 1425-1438

[51] Horacek B, Kim J, Kiger K. Spray cooling using multiple nozzles: Visualization and wall heat transfer measurements. IEEE Transactions on Device and Materials Reliability. 2004; **4**(4):614-625

[52] Newton E. Vapor chamber cooling finds growing role in hot products. Embedded; 30 May 2021. Available from: https://www.embedded.com/vapor-chamber-cooling-finds-growing-role-in-hot-products/ [Accessed: 3 August 2021]

[53] Grover GM, Cotter TP, Erickson GF. Structure of very high thermal conductance. Journal of Applied Physics. 1964;**35**:1990

[54] Glover G, Chen Y, Luo A, Chu H. Thin vapor chamber heat sink and embedded heat pipe heat sink performance evaluations. 25th Annual IEEE Semiconductor Thermal Measurement and Management Symposium, 2009, pp. 30-37. DOI: 10.1109/STHERM.2009.4810739

[55] Coolermastercorp: https://web2.coolermastercorp.com/wp-content/uploads/2018/01/ultra-slim-vapor-chamber-2.png

Chapter 5

Direct Contact Heat and Mass Exchanger for Heating, Cooling, Humidification, and Dehumidification

Marip Kum Ja, Qian Chen, Muhammad Burhan, Doskhan Ybyraiymkul, Muhammad Wakil Shahzad, Raid Alrowais and Kim Choon Ng

Abstract

A direct-contact heat and mass exchanger (DCHME) has many advantages over a traditional surface-type heat exchanger, including a high heat transfer coefficient, simplicity of design, and low OPEX and CAPEX. DCHME has a capability to exchange of both heat and mass between the two fluids in the same process. Hence, DCHMEs are widely used in numerous applications in various industries, including the air conditioning industry for cooling and dehumidification and heating and humidification. Based on their structure, DCHME can be categorized into two groups; two fluids direct contact (TFDC) exchanger and two direct contacts with one non-contact fluid (TDCONF) exchanger. This study developed a mathematical model for these two types of exchangers by using a discretized volume with distributed lumped-parameters method instead of using the conventional log mean enthalpy difference (LMHD) and NTU-effectiveness method. Thus, this model can reflect both heat and mass transfer behavior in every spatially distributed physical system. The objective of this study is to develop a mathematical model to be used as a tool for designing DCHME and to be applied as a sub-function of the model predictive control system to predict the effectiveness and dependent parameters of DCHME under the different load conditions and its various input parameters.

Keywords: heat and mass exchanger, distributed lump model, humidification and dehumidification

1. Introduction

For more than 100 years ago, direct contact heat and mass exchangers (DCHME) have been widely used in various industries, including chemical process plant, food, and beverage industry, geothermal heat recovery, seawater desalination, waste heat recovery, energy storage systems, production of steam generation for the Rankine power cycle, air conditioning and refrigeration industries, and many so forth. DCHME is a device in which the two process streams are flowing and contacting each other to exchange heat and mass between these two streams, which can be gas-solid, gas-liquid, liquid-liquid, liquid-solid, or solid-solid streams. The limitation of DCHME is the contamination of the streams depending on the degree of miscibility. Although it has a limitation, there are many advantages such as no corrode or foul or no degradation of heat transfer performance due to the lack of surfaces, a larger heat transfer surface area, much lower flow resistance compared with surface-type heat exchangers, and less capital and operational cost [1]. Thus, DCHMEs are widely used in air conditioning industry for cooling, heating and humidification, cooling and dehumidification, and cooling and humidification, such devices are swamp cooler or direct contact evaporative cooler, cooling tower, air washer spray chamber, cooling coil of air handling unit (AHU), direct expansion (DX) evaporator coil, indirect evaporative cooler, and M-cycle dew point evaporative cooler. Before developing the numerical model of these devices, the process of each device has required to analysis with the psychometric chart. There are total of eight basic air conditioning processes, which are plotted on a psychometric chart as shown in **Figure 1**. These processes are the air stream from the initial state O to the state (1) for sensible cooling, (2) for cooling and humidification, (3) for humidification only, (4) for heating and humidification, (5) for sensible heating, (6) for heating and dehumidification, (7) for dehumidification only, and (8) for cooling and dehumidification. In order to achieve the above eight processes, some processes need to transfer heat (sensible heat) only but some need to exchange heat and mass (sensible and latent heat) from the air stream. Process 1, cooling, and process 5, heating, are pure heat transfer processes only, which means removing heat from the air stream for cooling and adding heat to the air stream for heating without changing the moisture of the airflow. Process 3, humidification, and process 7, dehumidification, are pure mass transfer processes without variation in air temperature. The rest processes 2 (cooling and humidification), process 4 (heating and humidification), process 6 (heating and dehumidification), and process 8 (cooling and dehumidification) are both heat and mass (sensible and latent heat) transfer process that is adding or removing of heat and moisture to or from the air stream. For DCHME has a capability to transfer both heat and mass in the same process, almost all the processes except six can be accomplished by using DCHME in a single stage. The typical five DCHMEs mentioned above extensively used in the air conditioning industry are discussed with their basic air conditioning process.

(1) Swamp cooler or direct contact evaporative cooler, shown in **Figure 2(a)**, is a device in which the outside air flows through the medium of cooling pad or fill

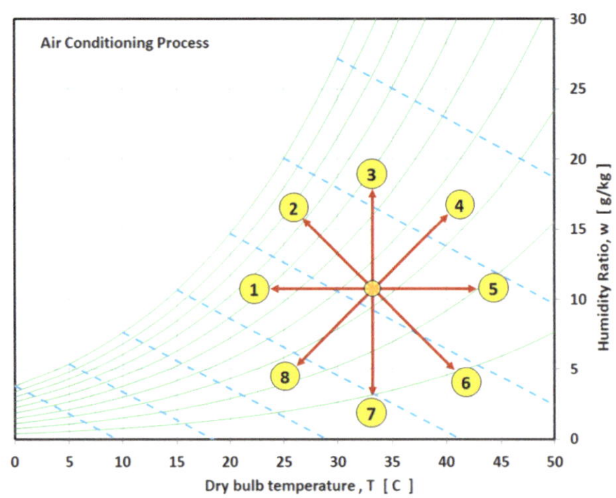

Figure 1.
Basic eight air conditioning processes plotted on the psychometric chart.

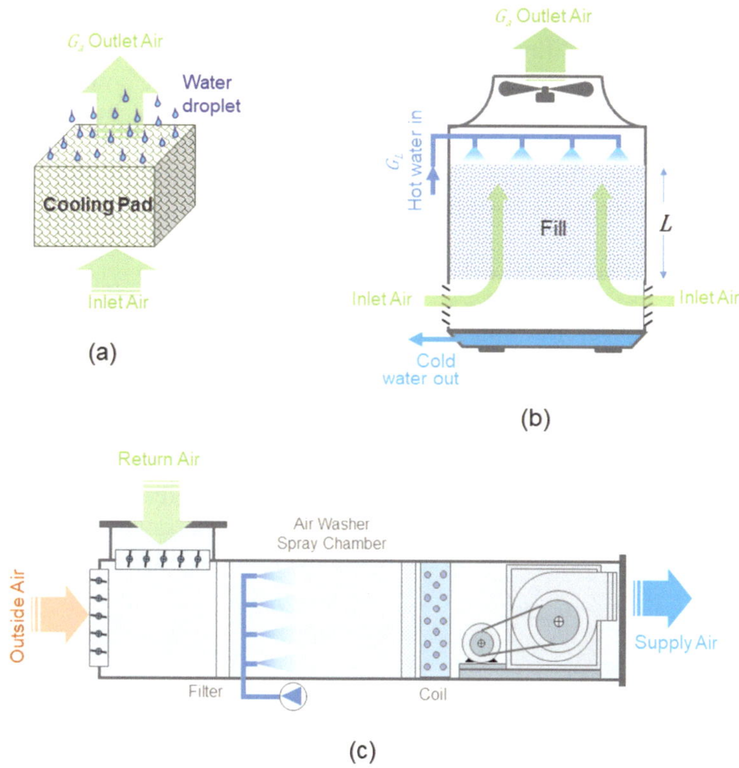

Figure 2.
Two fluids direct contact (TFDC) heat exchanger: (a) direct evaporative cooler/swamp cooler, (b) cooling tower, and (c) air washer chamber.

that is wetted with water by dripping or spraying. The air becomes cool and humidified (process 2) after passing through the medium that results from the absorption of heat from the air by the evaporation of water droplets or water film of the cool pad. For it is a passive cooling device, it has a very efficient electrical energy consumption, and so that, it is a very efficient space cooling in the hot and dry region with the lowest initial cost and operation cost.

(2) Cooling tower, shown in **Figure 2(b)**, is not an air conditioner device, but it is used extensively as a heat sink or heat removal from the condenser of the refrigeration circuit in the air conditioning industry. Likewise, in a swamp cooler, the hot water, carrying the rejection heat from the condenser, can be cool down by evaporating its water droplet or water film from the fill. In the cooling tower, the hot water is cooling down as the air stream is heated up and humidified (process 2) by transferring heat and mass between these two fluids.

(3) Air washer spray chamber, shown in **Figure 2(c)**, is a typical DCHME, which is widely used as an air conditioner and works in synergy with the heating or cooling coil in the air handling unit (AHU). The air washer chamber has the capability to condition the inlet air into three desired outlet air state-points by controlling the spraying water temperature. Heating and humidification, like process 4, can be achieved by spraying hot water into the chamber if its spray temperature is higher than the inlet air dry-bulb temperature. Cooling and dehumidification, like process 8, can be achieved by spraying chilled water which temperature is lower than the dew point temperature of the inlet air of the chamber. The inlet air can also be cooled as per process 2 line without changing the moisture content of air by spraying the chilled water temperature, which is lower than the inlet air dry-bulb temperature but higher than its dew point temperature.

Different structures but similar to another type of DCHME are plate finned tube heat exchanger used as a direct expansion (DX) evaporator coil, cooling coil operating with chilled water, or non-volatile refrigerant, shown in **Figure 3(a)**. Indirect evaporative cooler, shown in **Figure 3(b)**, and M-cycle dew point evaporative cooler, shown in **Figure 3(c)**, is also another type of DCHME. These types of exchangers have three fluid streams in which two streams directly contact each other, and the last stream is separated by the copper tube wall or separator sheet, shown in **Figure 3**.

(4) Plate finned tube heat exchangers are the most common type of heat exchanger used extensively in the air conditioning industry using as a DX evaporator coil and cooling coil operating with chilled water or ethylene/propylene glycol. For this case, if the dew point temperature of inlet air is lower than the surface temperature of the exchanger, the air stream's water vapor starts turning into condensate as a form of droplets or film on its surface. Therefore, depending on the dew point and surface temperature, the exchanger, can be thoroughly wetted, fully dry, or partially wetted. The fully dry or dry part of a partially wetted exchanger cannot be presumed as a heat and mass exchanger but as a conventional heat exchanger that can be calculated by its traditional LMDT or NTU-effectiveness

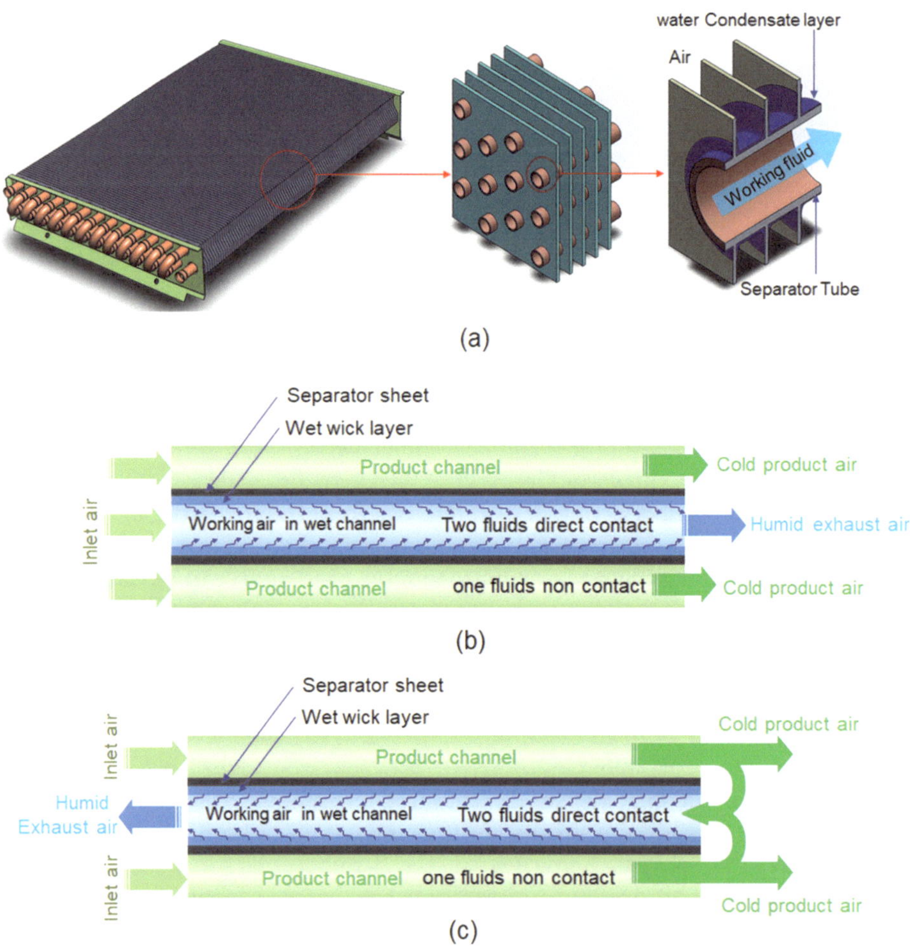

Figure 3.
Two direct contact with one non-contact fluids (TDCONF) heat and mass exchanger: (a) plate finned-tube heat exchanger operating with non-volatile or volatile working fluid, (b) indirect evaporative cooler, (c) M-cycle dew point evaporative cooler.

method. However, the fully wetted or wet part of the partially wetted exchanger is a direct contact heat and mass exchanger because heat and mass are exchanged between the two fluids of the air stream and the film of water condensate [2], shown in **Figure 3(a)**. The process rendering in the wet region of this exchanger is cooling and dehumidification of the air stream as process 8.

(5) Indirect evaporative cooler and M-cycle dew point evaporative cooler, shown in **Figure 3(b)** and **(c)**, are a combination of a direct evaporative cooler and air to the air heat exchanger. These coolers are composed of multiple duo wet and dry channels separated by a thin film sheet. The working air flowing through the wet channel is directly in contact with a thin water film of wick material or water mist sprayed into the wet channel for the evaporation process. The required heat for the evaporation is transferred from the product air flowing in the dry channel through the separator sheet. Thus, the product air temperature is cooling down along process 1 without changing the moisture content. For the indirect evaporative cooler, the working air at the beginning of the wet channel will be cooled and humidified like process 2 due to evaporation, but later part of the wet channel, the product air is heated and humidified due to the heat that transfers from the product air, like process 4. Likewise, indirect evaporative cooler, M-cycle dew point evaporative cooler has a similar process, but there has a concise process 2 at the beginning of the wet channel because the working air is partially taken from the product air [3–5].

Based on the number of fluids flowing in the exchanger and their arrangement, DCHME for air conditioning can be categorized into two main groups, which are shown in **Figure 4**. In the first group, the two fluids streams, water, and air are flowing in parallel or counter flow direction and directly contacting each other to exchange heat and mass between these two fluids. This type of exchanger is noted as Type-1 two fluids direct contact (TFDC) heat and mass exchanger. Examples of these exchangers are: (1) air washer chamber, (2) cooling tower, and (3) swamp cooler or direct contact evaporative cooler, shown in **Figure 2**. A single discretized element of this type of TFDC exchanger is shown in **Figure 5**.

The second group of DCHME has three kinds of working fluids in which two fluids, water, and air, are directly contacting each other to exchange heat and mass between these two fluids. The third fluid is separated by a tube wall or a thin sheet

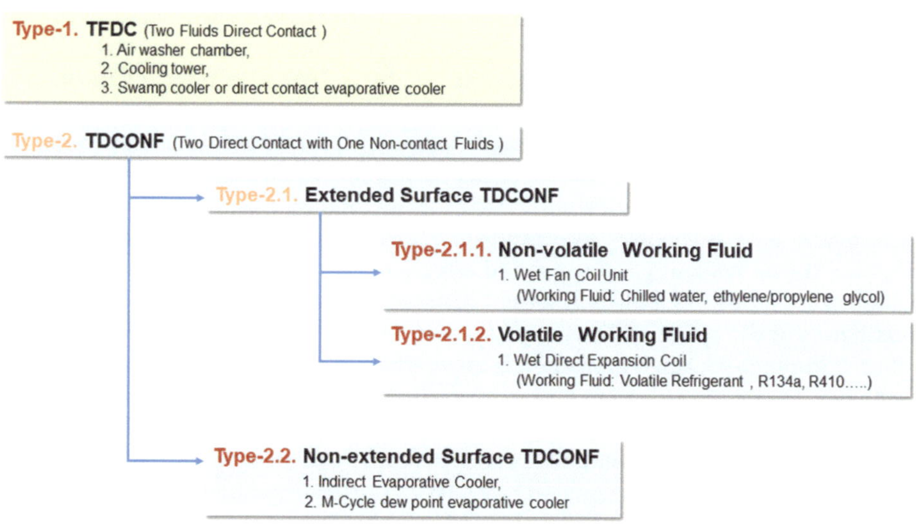

Figure 4.
Categorization of direct contact heat and mass exchangers that widely used in air-conditioning industry.

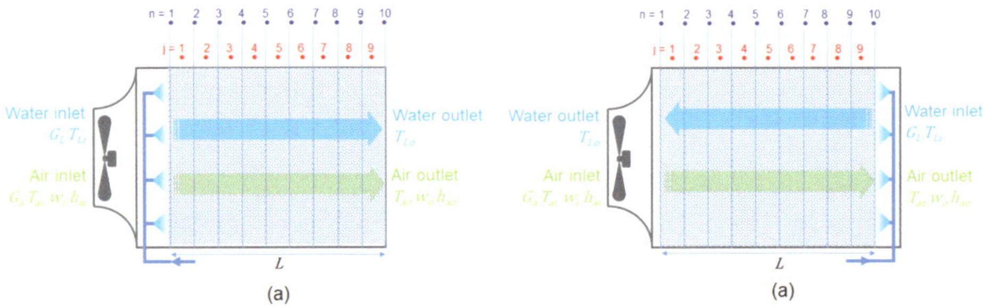

Figure 5.
TFDC exchanger with nine discretized elements, j = 9 *and* n = 10, *and water is spraying into a (a) parallel or (b) counter flow with air flow direction.*

layer to prevent mass transfer but only allow heat transfer. Examples of this type of heat and mass exchanger are shown in **Figure 3**, and it can be noted as Type-2, two direct contacts with one non-contact fluid (TDCONF) heat and mass exchanger. This TDCONF exchanger can also be divided into two groups based on the separator type. The first group, Type-2.1 has a separator with an extended surface. An example of an extended surface TDCONF exchanger is the wet region of the plate finned-tube cooling coil unit. Based on the nature of the working fluid, Type-2.1 can be sub-categorized into two groups. The first group, Type-2.1.1 of extended surface TDCONF exchanger, works with non-volatile refrigerants such as chilled water, ethylene/propylene glycol, etc., so that the working fluid temperature is changing continuously along with the coil rows depth. The second group, Type-2.1.2, works with volatile refrigerants such as R134a, R410, etc. Due to the evaporation in the working fluid flow, most of the exchanger has the same surface and fluid temperature profile. Although these two subgroups are the same exchanger, their mathematical models are different due to their different working fluid temperature profiles. Another group, Type-2.2 non-extended surface TDCONF exchangers, are using a plain sheet or tube as a separator, and examples of this type of exchangers are in the direct evaporative cooler and M-cycle dew point evaporative cooler. The mathematical model of this exchanger is the same with Type-2.1.1 of extended surface TDCONF exchanger except for the separator's area calculation.

Kays and London [6] introduced the definitions of effectiveness and NTU method to use in heat exchanger design in their 1955 publication. London et al. [7] used this method to fit the experimental data of the cooling tower, but this method is not generally consistent with all the other units [8]. The reason for the inconsistency is that the method was based only on the sensible heat transfer process, yet the mass exchanging process is excluded. Berman [9] introduced the log-mean enthalpy method (LMED) to reflect both the heat and mass transfer process of cooling tower design. Several studies used this method to analyze heat and mass exchangers, such as cooling towers and spray chambers [9–11]. LMED method can calculate the dependent parameters and effectiveness of the exchanger without using iteration process, but this method has some limitations to calculate all the parameters of the spatially distributed system. Thus, discretized volume with distributed lumped-parameters method is an alternative approach to develop the exchanger model with element nodes. This method is widely used to study the behavior of spatially distributed parameters of electrical systems [12], chemical reaction systems [13], heat and mass transfer [14, 15], microwave [16, 17], acoustics [18], and so many other systems. In this study, discretized volume with distributed lumped-parameters model was developed for both types of exchangers, Type-1 TFDC and Type-2 TDCONF, based on the principle of the graphical method mentioned in ASHRAE Fundamentals Handbook [19] and Systems and Equipment

Handbook [20]. The derivation of the model for Type-1 TFDC is explained in Section 3, and for Type-2 TDCONF exchanger is described in Section 4. The relation of convective heat transfer with mass transfer coefficients between air and water film, which are related to both models, is explained in Section 2.

2. The relation between convective heat and mass transfer coefficient

All types of DCHME mentioned above can be assumed that each exchanger has similar geometry and boundary conditions for heat and mass transfer process so convective mass transfer is analogous to convective heat transfer, and that can be applied for both laminar and turbulent flows [19]. Bird et al. [21] and Incropera and DeWitt [22] defined the analogy Nusselt (Nu) and Sherwood (Sh) number for the calculation of mass transfer coefficient h_M [m/s] from the heat transfer coefficient α_a [W/m^2.K]. Nusselt (Nu) number, related with α_a, is a function of Reynold (Re) and Pantanal (Pr) number, expressed in Eq. (1). Sherwood (Sh) number, related with h_M, is a function of Reynold (Re) and Schmidt (Sc) number, stated in Eq. (2).

$$Nu = \frac{\alpha_a\, D_H}{k_a} = a\,[Re]^b\,[Pr]^{1/3} \tag{1}$$

$$Sh = \frac{h_M\, D_H}{D_v} = a\,[Re]^b\,[Sc]^{1/3} \tag{2}$$

where a and b are coefficient numbers, D_H [m] is characteristic length of exchanger, k_a [W/m.K] is thermal conductivity of air, and D_v [m^2/s] is mass diffusivity. Division of Eq. (1) with Eq. (2) are called the Reynold analogy, and it gives the relation of Lewis (Le) number which can be seen in Eqs. (5) and (6).

$$\frac{\alpha_a}{h_M} = \left[\frac{Pr}{Sc}\right]^{\frac{1}{3}} \frac{k_a}{D_v} \tag{3}$$

$$\frac{\alpha_a}{h_M} = \left[\frac{Pr}{Sc}\right]^{\frac{1}{3}} \frac{Cp\,\mu}{Pr}\,\frac{\rho\,Sc}{\mu} = Le^{-\frac{1}{3}}\,Le\,\rho\,Cp_m \tag{4}$$

$$\frac{\alpha_a}{h_M \rho\, Cp_m} = Le^{\frac{2}{3}} \approx 1 \tag{5}$$

where μ [N s/m^2] is dynamic viscosity, Cp_m [J/kg.K] is specific heat capacity of moist air, and ρ [kg/m^3] is density of air. For the humidity ratio is taken as a driving force in mass transfer process, mass transfer coefficient should be defied with the term $K_M = h_M\,\rho$ [kg/m^2.s]. The Lewis relation equation Eq. (5) can be written as follows:

$$\frac{\alpha_a}{K_M Cp_m} = Le^{\frac{2}{3}} \approx 1 \tag{6}$$

Similarly, Chilton and Colburn [23] proposed Chilton-Colburn j-factor analogy using this similarity to relate Nusselt number to friction factor by the analogy. The j-factor analogy has some limitation that is it can be only reliable when the surface conditions are identical. Bedingfield and Drew [24] also proposed the relation equation between heat and mass coefficients. Many others relation equations that can be found in literatures to calculate mass transfer coefficient h_M from heat transfer coefficient α_a, but in this study, Lewis relation will use in this model development due to its simplicity.

3. Type-1: two fluids direct contact (TFDC) exchanger

TFDC heat exchangers, such as air washer chamber, cooling tower, and swamp cooler or direct contact evaporative cooler, are the first type of DCHME shown in **Figure 2**. TFDC has two water and air fluids flowing in parallel or counter, and directly contacting each other to exchange heat and mass between them. **Figure 6(a)** shows the basic one discretized element of air washer spray chamber and direct contact evaporative cooler. The air stream is directly in contact with the saturated layer of each water droplet or the fill/cool pad, and heat and mass are transferring between them according to the mass transfer coefficient K_M with mass transfer area a_M and heat transfer coefficient α_a with heat transfer area a_H. **Figure 6(b)** shows the fundamental discretized element of the cooling tower and swamp cooler. Both elements have the same concept of numbering node for three layers parameters; air and its average, water and its average, and saturated layer parameters, shown in **Figure 6(c)**.

3.1 Energy balance between two fluids of Type-1 model (energy balance line, EBL)

In TFDC exchanger, the water is sprayed into the chamber along the airflow (parallel flow), shown in **Figure 5(a)**, or against the airflow (counter flow), shown in **Figure 5(b)**. Due to the difficulty of area measurement, TFDC exchangers are discretized in length along the fluids flowing, assuming it has same cross sectional area. **Figure 5** shows that the system is divided into nine differential volumes numbered by "*j*," and the inlet and outlet of the differential volumes are noted by "*n*." Since each differential volumes has the same cross sectional area, the differential volume can be changed with differential length "*dL*." and mass flowrates of water and air are needed to be changed to mass flux or flow rate per unit cross-sectional area of the exchanger, A_{CS} [m^2], for air G_a [kg/(s·m^2)], and water G_L [kg/(s·m^2)].

According to the conservation of energy, the energy transferred from one fluid is equal to the energy gain of the other fluid. Assuming that heat loss from the

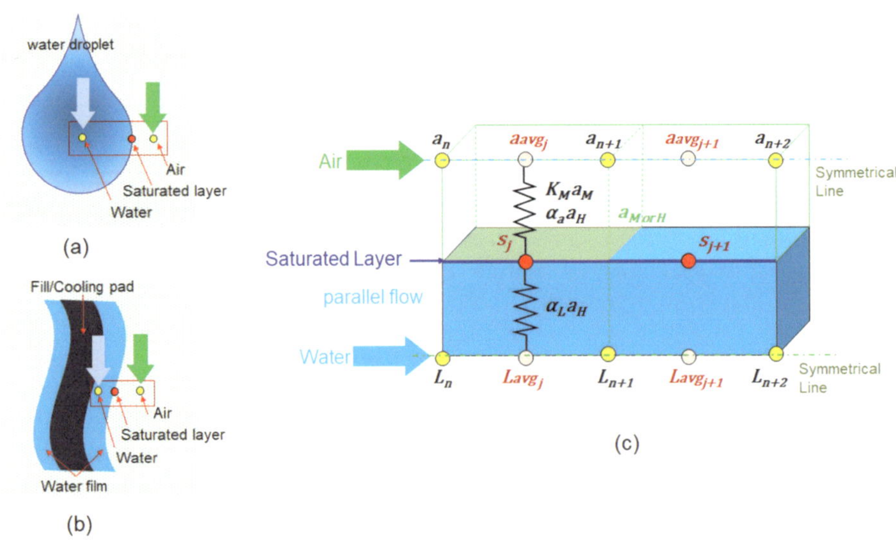

Figure 6.
(a) Single discretized element of the air washer spray chamber or direct contact evaporative cooler, and (b) of cooling tower or swamp cooler, and (c) node numbering of the points for air stream, water flow, and saturated layer of each element.

system and the change of air or water flow rate due to evaporation or condensation are comparatively small and negligible. Hence, the energy equation of two fluids can be expressed as follows:

$$G_a dh_a = -G_l Cp_l dT_l \tag{7}$$

where, dh_a [kJ/kg] and dT_l [°C] are the enthalpy and temperature difference between outlet and inlet of one discretized element, and Eq. (7) can be rewritten with element node number for the parallel and counter flow as follows:

$$\frac{h_a^{n+1} - h_a^n}{T_l^{n+1} - T_l^n} = -\frac{G_l Cp_l}{G_a} \tag{8}$$

$$\frac{h_a^{n+1} - h_a^n}{T_l^n - T_l^{n+1}} = -\frac{G_l Cp_l}{G_a} \tag{9}$$

The generalized energy balance line equation with node number for the counter and parallel flow can be described as follows:

$$\frac{h_a^{n+1} - h_a^n}{T_l^{n+1} - T_l^n} = \pm\frac{G_l Cp_l}{G_a} \tag{10}$$

A minus sign refers to parallel flow of air and water stream. A plus sign refers to counter flow (water flow is in the opposite direction of airflow).

3.2 Heat and mass transfer between two fluids of Type-1 model (tie-line slope, TLS)

According to the conservation of mass between the two fluids, change of water amount in liquid stream, dG_l, due to evaporation (if saturated humidity ratio, w_s [kg/kg], at the saturated layer is greater than humidity ratio, w_a, of air stream) or condensation (if $w_s < w_a$) is equal to the amount of humidification or dehumidification of air stream, $-G_a d_w$, in the spray chamber. The above statement can be expressed by equation as follows:

$$\Delta mass = dG_l = -G_a dw_a \tag{11}$$

The negative sign means that the increasing of mass in one fluid stream is equal to the losing mass of other fluid stream. Similarly, the change of mass flow rate, $\Delta mass$ from either stream is equal to the mass flux [kg/m^2], that is, mass transfer from the air stream to the saturated liquid surface (or) vise visa per contact surface area along the exchanger, A_{CT} [m^2]. It can be written in equation as follows:

$$\Delta mass = K_M a_M \left(w_s - w_{aavg}\right) dL \tag{12}$$

where, K_M [kg/(s·m^2)] is mass transfer coefficient, that is, mass transfer rate per contact surface area along the exchanger, A_{CT} [m^2]. a_M [m^2/m^3] is a mass transferring contact surface area between air and water per unit volume of spray chamber. L [m] is the chamber length. The sensible heat transfer between air stream and saturated liquid surface can be expressed as follows:

$$G_a C_{pm} dT_a = \alpha_a a_H \left(T_s - T_{aavg}\right) dL \tag{13}$$

where, α_a [W/(m^2·K)] is convective heat transfer coefficient of air. a_H [m^2/m^3] is heat transfer contact surface area between air and water per unit volume of spray chamber. C_{pm} [kJ/(kgda·K)] is a specific heat of moist air at constant pressure. T_s [°C] is saturation temperature of water film. T_{aavg} [°C] is the average air temperature of two adjacent nodes, see in **Figure 6(c)**. The total heat transfer between air stream and saturated liquid surface is equal to the sum of sensible heat, $Cp_m dT_a$, and latent heat, $h_{fg}dw$. The total heat transfer can be calculated by combining of mass transfer Eq. (12) multiplying with enthalpy of vaporization h_{fg} [kJ/kg] and the sensible heat transfer Eq. (13). The total heat transfer equation can be expressed as follows:

$$G_a\left(Cp_m dT_a + h_{fg}dw\right) = \left[K_M a_M\left(w_s - w_{avg}\right)h_{fg} + \alpha_a a_H\left(T_s - T_{aavg}\right)\right]dL \quad (14)$$

Since $Cp_m dT_a + h_{fg}dw$ is equal with enthalpy change of air dh_a, and a small change of water vaporization heat with the function of temperature is neglected, the Eq. (14) can be rewritten as follows:

$$G_a dh_a = K_M a_M\left[\left(w_s - w_{avg}\right)h_{fg} + \frac{\alpha_a a_H}{K_M a_M}\left(T_s - T_{aavg}\right)\right]dL \quad (15)$$

The contact surface areas per system volume, a_H and a_M for the heat and mass transferring process in spray chambers are identical ($a_H = a_M$), however, this hypothetical is not always correct, especially, if the systems are using the packing materials as an extension of contact surface area which are not fully wetted. Assuming Lewis number with the power of 2/3 is similar with 1, see in Eq. (6), and neglecting of a small variations in h_{fg}. McElgin and Wiley [25] simplified Eq. (15) as follows:

$$G_a dh_a = K_M a_M\left(h_s - h_{aavg}\right)dL \quad (16)$$

The sensible heat transferring from saturated liquid surface to the working fluid stream is equal to the heat gain of the working fluid, the energy balance equations can be depicted as follows:

$$G_l Cp_l dT_l = \alpha_l a_H\left(T_s - T_{lavg}\right)dL \quad (17)$$

After substitution of Eqs. (16) and (17) into Eq. (7), and rearranged, the Eq. (7) can be written as follows:

$$-\alpha_l a_H\left(T_s - T_{lavg}\right)dL = K_M a_M\left(h_s - h_{aavg}\right)dL \quad (18)$$

$$\frac{h_s - h_{aavg}}{T_s - T_{lavg}} = -\frac{\alpha_l a_H}{K_M a_M} = -\frac{\alpha_l\, Cp_m}{\alpha_a} \quad (19)$$

According to the above Eq. (19), the value of, $-\frac{\alpha_l\, Cp_m}{\alpha_a}$, is a slope or gradient line of the rise, the enthalpy incensement in vertical over the run, temperature incensement along the horizontal in T-h graph. In other words, the slope is a ratio of the enthalpy difference between air and saturated surface to the temperature difference between working fluid and saturated surface. This slope, the relation of the temperature and enthalpy, can be denoted as tie-line slope, *TLS* of the Type-1 TFDC exchanger. The *TLS*, Eq. (19), can be solved by using graphical method, drawing in water-air T-h graph which is proposed in [2, 19, 20, 26], but this method cannot be applied in the numerical model. In order to solve TLS equation numerically, the two unknown parameters, saturated temperature, T_s, and saturated enthalpy, h_s, have to reduce one unknown parameter by substitution of their relation equation that can be obtained by fitting of the saturation line on T-h graph, calculated from the psychometric chart equations depicted in ASHRAE code [19], with the third order polynomial regression equation, Eq. (20).

$$h_s = coe1.T_s^3 + coe2.T_s^2 + coe3.T_s + coe4 \tag{20}$$

Figure 7 shows the third order polynomial regression line with the saturation point, in red circle legend, calculated by psychometric chart equations, and its four related coefficients values, *coe*1 to *coe*4. The regression has a very high accuracy as its R-square value 0.9993706, which covering the range of air temperature from 0 to 50°C.

By the substitution of polynomial regression equation Eq. (20) into Eq. (19), the tie-line slope equation become one unknown parameter equation, and that can be rewritten with node number n and j as follows:

$$coe1\left(T_s^j\right)^3 + coe2\left(T_s^j\right)^2 + (coe3 + TLS)T_s^j + coe4 - \left(\frac{h_a^n + h_a^{n+1}}{2}\right) - TLS\left(\frac{T_l^n + T_l^{n+1}}{2}\right) = 0 \tag{21}$$

Newton Raphson iteration method or any other relevant numerical method can be applied to solve out the unknown root value T_s of each node j of the equation, Eq. (21), by using the known value of T_l and h_a of each node n (node numbering of each element is shown in **Figure 6**).

3.3 Dry bulb temperature of each node for Type-1 model

The last unknown parameter of air temperature, T_a [°C], of each node n, can be calculated from the ratio of sensible heat Eq. (13) to total heat and mass transfer Eq. (16), and it gives as follows:

$$\frac{dT_a}{dh_a} = \frac{\alpha_a a_H \left(T_s - T_{aavg}\right)}{K_M C_{pm} a_M \left(h_s - h_{aavg}\right)} \tag{22}$$

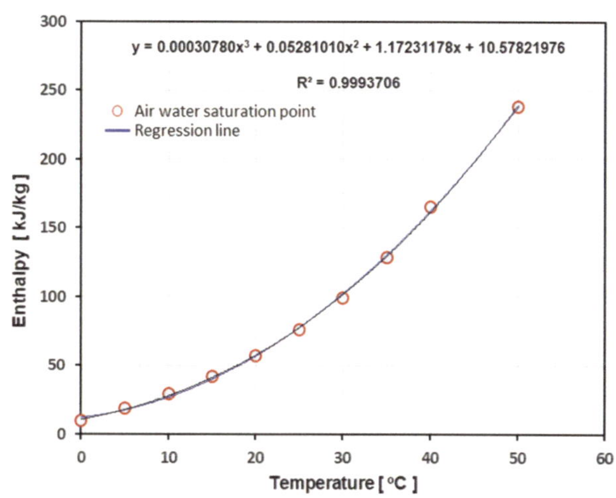

Figure 7.
Third order polynomial regression equation of the enthalpy of the saturated air.

The value of $\frac{\alpha_a}{K_M C_{pm}}$ can be taken as 1 because the Lewis number with the power of 2/3 is approximately equal to 1, see in Eq. (6). For the TFDC exchanger, this study assumes that the contact surface area per volume of heat transferring process, a_H, is identical with the one of mass transferring process, a_M. For numerical calculation, the average parameters, T_{aavg} and h_{aavg}, are located between two adjacent n nodes, and noted with j node, shown in **Figure 6**. After rearranging of the Eq. (22), it gives with node number n and j as follows:

$$T_a^{n+1} - T_a^n = \frac{\frac{h_a^{n+1} - h_a^n}{\left(h_s^j - h_{aavg}^j\right)}\left(T_s^j - T_a^n\right)}{\left(1 + \frac{h_a^{n+1} - h_a^n}{2\left(h_s^j - h_{aavg}^j\right)}\right)} \tag{23}$$

The outlet air temperature of each node, T_a^{n+1}, can be calculated from the known inlet temperature, T_a^n, air enthalpy, h_a, average air enthalpy, h_{aavg}, vapor saturated air enthalpy, h_s, and vapor saturated air temperature, T_s by using Eq. (23).

3.4 Contact length calculation for Type-1 TFDEC exchanger model

Calculation of the contact length or depth of packing of the exchanger, L [m], is a primary interest for the designing of the system to achieve a desired out let condition of air or out let water temperature. The length can be calculated by the integration of heat and mass transfer equation, Eq. (16).

$$L = \frac{G_a}{K_M a_M} \int_{j=1}^{j=N-1} \frac{1}{\left(h_s - h_{aavg}\right)} dh_a \tag{24}$$

The integral equation, Eq. (24) can be solved by using Sampson's rule or Trapezoidal integration method. The integral equation can be rearranged and rewritten with node number n and j as follows:

$$L = \frac{G_a}{K_M a_M} \sum_{j=1}^{j=N-1} \left[\frac{1}{\left(h_s^j - h_{aavg}^j\right)} \times \left(h_a^{n+1} - h_a^n\right)\right] \tag{25}$$

Change of the air mass flow rate, G_a, due to the transfer of water vapor from the water droplet or film to the air by evaporation or condensation is very minimal compared with its mass flowrate, and it can be taken as negligible for all the practical applications. The contact length or exchanger length L, can be calculated from the known air enthalpy, h_a, and saturated enthalpy, h_s, of each elements.

4. Type-2: two direct contact with one non-contact fluids (TDCONF) exchanger

Section 1, discussed about the DCHME that widely used in air conditioning industry, has categorized the exchangers into two groups. Mathematical model of the first type TFDC heat exchanger has been discussed in Section 3. The development of mathematical model for the second type of TDCONF exchanger will be discussed in detail in this section. The examples of TDCONF exchangers are the wet region of plate finned tube heat exchanger for the direct expansion (DX) evaporator coil, cooling coil unit working with chilled water or non-volatile refrigerant, indirect evaporative cooler, and M-cycle dew point evaporative cooler, which are shown in **Figure 3**.

A discretized volume with distributed lump parameters model of TDCONF, shown in **Figure 8**, has three fluids, non-contact fluid, contact fluid, and wetted wick layer or thin water film. The product air of indirect evaporative cooler and volatile refrigerant or chilled water of the wetted plate finned tube cooling coil are the examples of non-contact fluid. These fluids are separated by the separator sheet or tube from thin water film and contact fluid air to prevent the transferring of mass and only to allow the transferring of heat between non-contact fluid and thin water film. Thermal resistance due to the conduction of separator sheet or metal tube is R_{mw}, and due to the convective heat transfer in the non-contact fluid is R_l. But, both heat and mass are transferring between the thin water film and contact fluid air. Heat and mass transfer circuit between from the symmetrical line of contact fluid air to the non-contact fluid with the coefficient of mass transfer, K_M, and thermal resistances are shown in **Figure 8(c)**. Each discretized volume of all TDCONF exchanger is composed with three node points locating in line on the four layers of materials. The first node point, L_{avgj} is at the symmetrical line of non-contact fluid, s_j is at the surface of saturated water film, and a_{avgj} is at the symmetrical line of direct contact fluid air. This model assumed that the temperature difference between outer surface of separator sheet or copper tube and saturated surface of water film is negligible. TDCONF exchanger can be divided into three types of different numerical models: (1) Type-2.1.1 extended surface TDCONF exchanger working with non-volatile working fluid, (2) Type-2.1.2 extended surface TDCONF exchanger working with volatile refrigerant, and (3) Type-2.2 non-extended surface TDCONF exchangers, shown in **Figure 4**.

4.1 Type-2.1: Extended surface TDCONF exchanger

In air conditioning process, plate finned tube heat exchange are widely used for cooling and dehumidification process in fan coil unit (FCU) and air handling unit (AHU). If there is no condensation on the outer surface of finned tube heat exchanger, it is a normal heat exchanger and can be calculated by using conventional heat exchanger equation, UA-LMTD or NTU-effectiveness method.

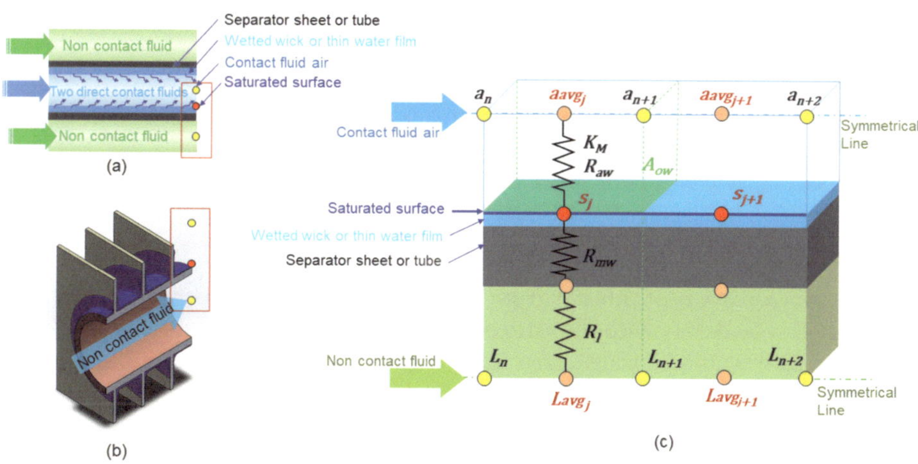

Figure 8.
A discretized volume with distributed lump parameters model of a typical (a) indirect evaporative cooler, (b) wetted region of plate finned tube cooling coil, and (c) node numbering of non-contact fluid (product air, volatile refrigerant, and chilled water), water saturated surface layer, separator (sheet or tube), and contact fluid (air).

However, if the surface temperature of heat exchanger is lower than the dew point temperature of the direct contact fluid air, the condensation process will be taken place and not only heat but also mass will be transferred between the air flow and the water condensate film. Hence, wet region of cooling heat exchanger can be assumed as a direct contact heat and mass exchanger.

4.2 Fin efficiency of extended surface TDCONF heat exchanger

In plate finned tube heat exchanger, fin efficiency is one of the critical parameters for the development of exchanger model, and that can be calculated from the physical dimensions of the heat exchanger. Equations (26)–(33) are the equations to calculate the physical parameters of heat exchanger; number of fins N_f [−], coil face or frontal area A_a [m^2], external exposed prime surface area A_p [m^2], external secondary surface area A_s [m^2], internal surface area A_i [m^2], total external surface area A_o [m^2], the ratio of external to internal surface area Br [−], and coil core surface parameters F_s [−] from its dimensions, width L_W [m], height L_H [m], depth L_D [m], fin gap f_g [m], fin thickness f_t [m], outside d_o [m], inside diameter d_i [m] of tube, longitudinal tube spacing S_L [m], and transverse tube spacing S_T [m]. This equations are based on heat exchanger with the continuous plate fin and tubes in stagger arrangement. For the other type of heat exchanger, their related equations are stated comprehensively in AHRI standard 410 [27], and that can be applied in the fin efficiency calculation.

$$N_f = \frac{L_W}{f_g} \tag{26}$$

$$A_a = \frac{L_H L_W}{1000000} \tag{27}$$

$$A_s = N_f \left[\frac{L_H L_D}{500000} - \frac{d_o^2 N_t}{636688} \right] \tag{28}$$

$$A_p = \frac{d_o L_W N_t - d_o t_f N_t N_f}{318344} \tag{29}$$

$$A_i = \frac{d_i N_t L_W}{318344} \tag{30}$$

$$A_o = A_s + A_p \tag{31}$$

$$Br = \frac{A_i}{A_o} \tag{32}$$

$$F_s = \frac{A_o}{A_f N_r} \tag{33}$$

Fins can increase heat transfer from a prime surface of heat exchanger. Fin efficiency can be define with the ratio of the actual heat transferred from the fin to the heat that would be transferred if the entire fin were at its root or base temperature [20]. Fin efficiency equation can be written as follows:

$$\varnothing_f = \frac{\int \alpha_a (T_f - T_a) dA}{\int \alpha_a (T_{fr} - T_a) dA} \tag{34}$$

where $\varnothing_f$ is the fin efficiency, T_a is the temperature of the surrounding environment, and T_{fr} is the temperature at the fin root. T_f is the temperature along the fin. Fin efficiency decreases as the heat transfer coefficient increases because of the increased heat flow. Total heat transfer from a finned tube heat exchanger is sum of heat transfer from finned surface or secondary area A_s and un-finned or prime area A_p. It can write in equation as follows:

$$Q = (\alpha_p A_p + \varnothing_f \alpha_s A_s)(T_{fr} - T_a) \tag{35}$$

Assuming the heat transfer coefficients for the finned surface α_s and prime surface α_p are equal and note as α_a air side heat transfer coefficient, Eq. (35) can be rearranged as follows:

$$\varnothing_s = \left(1 - (1 - \varnothing_f)\frac{A_s}{A_o}\right) = \frac{A_p + \varnothing_f A_s}{A_o} \tag{36}$$

where A_o is total surface area of $(A_s + A_p)$. $\varnothing_s$ is fin effectiveness. Schmidt [28] presents empirical equation of fin surface efficiency for circular, rectangular, and hexagonal fins using an equivalent circular fin radius.

$$\varnothing_f = \frac{\tanh(m_p r_o \varphi)}{m_p r_o \varphi} \tag{37}$$

$$m_p = \sqrt{\frac{2\,\alpha_a}{k_a t_f}} \tag{38}$$

$$\varphi = \left[\left(\frac{r_e}{r_o}\right) - 1\right]\left[1 + 0.35\ln\left(\frac{r_e}{r_o}\right)\right] \tag{39}$$

where r_o is the outside tube radius, r_e is the equivalent circular fin radius, m_p is the standard extended surface parameter, φ is dimensionless thermal resistance. α_a is convective heat transfer coefficient of air. k_a is the thermal conductivity of the fin and t_{fin} is the fin thickness. Plate finned tube heat exchanger for stagger tube array can be considered as the integration of hexagonal fins. For hexagonal fins:

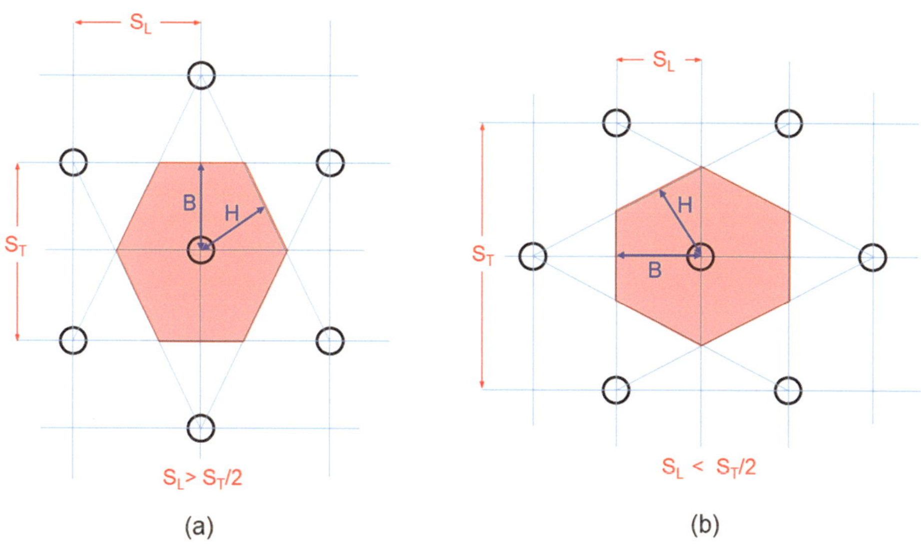

Figure 9.
Staggering arrangement of tube in plate finned tube heat exchanger (a) a half of transverse distance is lesser than longitudinal distance and (b) longitudinal distance is lesser than half of transverse distance.

$$\frac{r_e}{r_o} = 1.27\,\psi\sqrt{\beta - 0.3} \tag{40}$$

where Ψ and β are defined as; $\psi = B/r_o$ and $\beta = H/B$, β must be greater than 1. Depending on the *ST* (the transverse vertical tube spacing) and *SL* (the longitudinal horizontal tube spacing), shown in **Figure 9**, *B* and *H* can be defined as follows:

If $SL > ST/2$, then $B = ST/2$. If $SL < ST/2$, then $B = SL$.

$$H = \frac{1}{2}\sqrt{\left(\frac{S_T}{2}\right)^2 + S_L{}^2} \tag{41}$$

However, Schmidt's empirical equation of fin surface efficiency is limited to the situations of where $\beta > 1$.

When calculating the overall area of Type-2.1 extended surface TDCONF exchanger, the fin effectiveness term, $\varnothing_s$, can used to simplify the thermal resistance equation of heat and mass transfer from air to working fluid.

4.3 Type-2.1.1 extended surface TDCONF exchanger operating with non-volatile working fluid

Type-2.1.1 extended surface TDCONF exchanger is widely used as a cooling coil of air conditioning unit that operating with non-volatile working fluid such as chilled water or ethylene/propylene glycol. The air entering to the first part of the cooling coil is start decreasing its temperature but without moisture removing, this is called sensible heat removing part or dry coil region. When the air temperature is decreasing and reaching the dew point temperature of the entering air, moisture in the air stream start condensing on the surface of the coil called sensible and latent heat removing part or wet coil region.

4.3.1 Resistance of heat and mass transfer and dry-wet boundary region

Depending upon the dew point temperature of inlet air and working fluid temperature, the surface of the cooling coil can be fully dry, fully wet, or partially wet. The driving force for heat transfer in dry region is the temperature difference between air temperature T_a and working fluid temperature T_l, crossing the thermal resistance R_{ad}, R_{md}, and R_l. In the wet region, there are two types of heat and mass transfer processes synergy in series from the air to the working fluid. The first process is the exchange of heat and mass between the air and the saturated surface of water condensate layer due to their potential difference between h_a and h_s. The second process is transferring heat alone from the surface to the working fluid, and their driving force is the temperature difference between T_s and T_l. **Figure 10** shows a typical plate finned tube heat exchanger with counter flow process and thermal diagram of dry and wet region with their potential difference and resistances.

The capacity of the sensible heat Q_D that transferring in the dry region can be calculated by using the following equation:

$$Q_{Td} = U_D A_{oD} \Delta T_m = \frac{A_{oD} \Delta T_m}{R_{TD}} \tag{42}$$

where A_{oD} [m^2] is total surface area of dry region, R_{TD} [m^2.K/W] is total thermal resistance of dry region, and ΔT_m is log mean temperature difference between air dry bulb temperature and working fluid temperature. In wet region, total heat

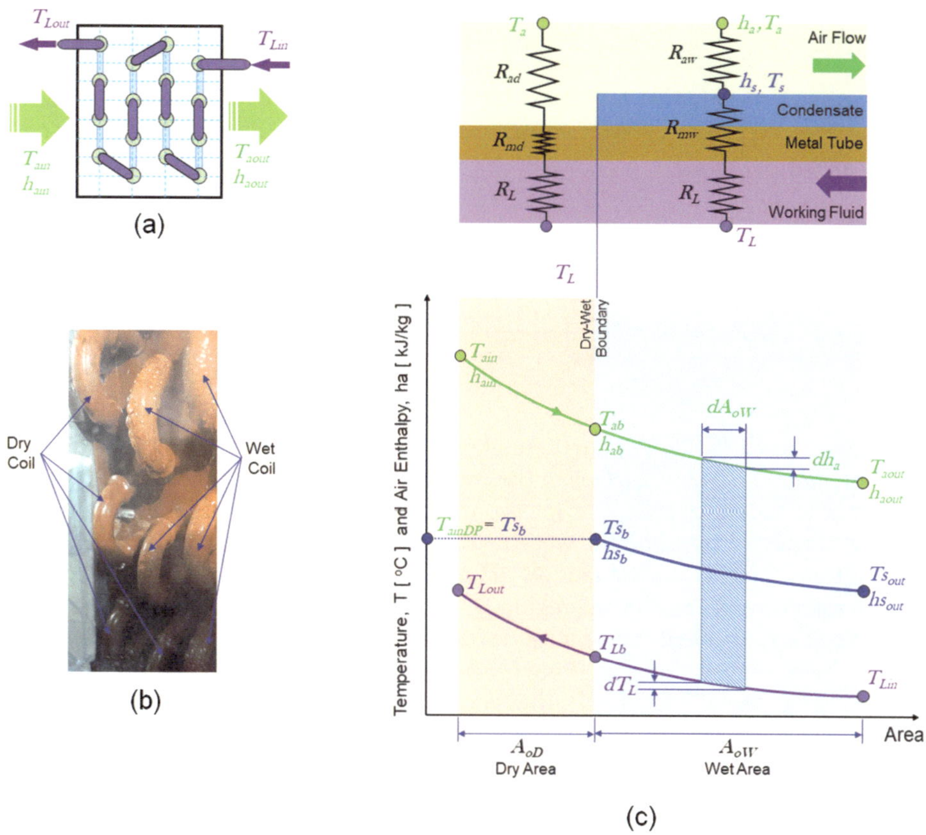

Figure 10.
(a) A typical single circuit with four rows deep coil plate finned tube heat exchanger with counter flow arrangement between air and working fluid, (b) photo record of dry and wet region, and (c) thermal diagram of dry and wet region of extended surface TDCONF exchanger and their potential difference of heat and mass transfer process.

capacity, Q_W, sensible heat and latent heat, transferred from air to saturated surface of condensate water can be calculated by using the following equation:

$$Q_{Tw} = U_W A_{oW} \Delta h_m = \frac{A_{oW} \Delta h_m}{Cp_m \, R_{aW}} \tag{43}$$

where A_{oW} [m^2] is total surface area of wet region, R_{aW} [m^2.K/W] is thermal resistance of air side in wet region, and Δh_m is log mean enthalpy difference between air and saturated wetted surface layer. For the calculation of heat transfer from air to working fluid, total overall thermal resistance of finned tube heat exchanger is required to calculate from the heat exchanger parameters and heat transfer coefficient. Overall thermal resistance with clean non-fouled surfaces for dry and wet cooling coils is a combination of three individual thermal resistances: (1) R_a [m^2.K/W] convective thermal resistance between air and external surface of the coil, (2) R_m [m^2.K/W] total conductance heat transfer of the metal fin, R_f, and tube, R_t, and (3) R_l [m^2.K/W] convective thermal resistance between the internal surface of the coil and the working fluid flowing inside the coil. Brown [29] derived thermal resistance of dry region, R_{md}, and wet region, R_{mw}, based on the heat transfer concept, proposed by Ware-Hacha [30], as follows:

$$R_{md} = R_f + R_t = \frac{(1 - \varnothing_s)}{\varnothing_s} R_{ad} + \frac{Br \, D_i \, Ln\left(D_o/D_i\right)}{2\, k_t} \tag{44}$$

$$R_{mw} = R_f + R_t = \frac{(1-\varnothing_s)}{\varnothing_s}\left(R_{aw}\frac{Cp_m}{m''}\right) + \frac{Br\, D_i\, Ln\left(D_o/D_i\right)}{2\,k_t}. \tag{45}$$

where m'' is $\frac{dh_s}{dT_s}$. The overall thermal resistance of dry R_{Td} and wet R_{Tw} region can be calculated as follows:

$$R_{Td} = R_{ad} + R_{md} + R_l = \frac{1}{\varnothing_s\,\alpha_{ad}} + \frac{Br\, D_i\, Ln\left(D_o/D_i\right)}{2\,k_t} + \frac{Br}{\alpha_l}. \tag{46}$$

$$R_{Tw} = R_{aw} + R_{mw} + R_l = \frac{1}{\alpha_{aw}} + \frac{(1-\varnothing_s)}{\varnothing_s}\left(\frac{1}{\alpha_{aw}}\frac{Cp_m}{m''}\right) + \frac{Br\, D_i\, Ln\left(D_o/D_i\right)}{2\,k_t} + \frac{Br}{\alpha_l}. \tag{47}$$

where R_{ad}, R_{aw} [m^2.K/W] are thermal resistance for convective heat transfer between air and external surface of the dry and wet coil, R_t [m^2.K/W] is thermal resistance of the metal tube, and R_l [m^2.K/W] is thermal resistance between the internal surface of tube and the working fluid flowing inside the coil. Br [−] is the ratio of total external surface area A_o to internal surface area A_i.

The ratio of wet and dry coil region is influenced by the characteristic of coil such as coil arrangement, refrigerant distribution, coil depth, entering air dew point and dry bulb temperatures, working fluid temperature, and their flowrates. Another parameter, Y [kg °C/kJ], the ratio of fluid temperature rise to air enthalpy drop, is also important parameter to calculate the boundary region. Y can be driven from the ratio of the heat that gained by working fluid to the heat that lost from the air stream. The equation is as follows:

$$\frac{dT_l}{dh_a} = \frac{m_a}{m_l Cp_l} = Y \tag{48}$$

The total heat transferring from the air to the saturated water surface in wet region of TDCONF exchanger is similar with TFDC exchanger, Eq. (16). Mass transfer coefficient KM is substituted with Eq. (6), and it can be written as follows:

$$dq_{Tw} = K_M(h_a - h_s)\, dA_{ow} = \frac{(h_a - h_s)\, dA_{ow}}{Cp_m R_{aw}} \tag{49}$$

Total heat transferred capacity dq_{Tw} in wet region also equals with the heat that transfer from the saturated wet surface layer to the working fluid that flowing inside the tube.

$$dq_{Tw} = \frac{(T_s - T_l)\, dA_{ow}}{R_{mw} + R_l} \tag{50}$$

After equating and rearranging of Eqs. (49) and (50), the following equation will be obtained, and can be denoted as the characteristic of the coil, C [2, 19, 20, 26].

$$C = \frac{R_{mw} + R_l}{Cp_m R_{aw}} = \frac{(T_s - T_l)}{(h_a - h_s)} \tag{51}$$

For the dry-wet boundary line, the coil characteristic C can be expressed as follows:

$$C = \frac{(T_{dpai} - T_{lb})}{(h_{ab} - h_{sb})} \tag{52}$$

Enthalpy of air at dry-wet boundary line, h_{ab} [kJ/kg] can be derived from Eqs. (48) and (52) as follows:

$$h_{ab} = \frac{T_{dpai} - T_{lo} + Y\,h_{ai} + C\,h_{sb}}{C + Y} \tag{53}$$

The exchanger coil can be determined whether it is fully wetted, fully dry, or partially wetted by comparing the values of h_{ab} with h_{ai}, and h_{ao}. If the value h_{ab} is higher than h_{ai}, all the surface of exchanger is fully wetted. If the value of h_{ab} is higher than h_{ao} but less than h_{ai}, the surface of exchanger is partially wetted. If the value of h_{ab} is less than h_{ao}, then the surface of exchanger is completely dry. The temperature of air at dry-wet boundary point, T_{ab} [°C], can be calculated from the heat energy equation of the air stream that flowing in dry region.

$$T_{ab} = T_{ai} - \frac{(h_{ai} - h_{ab})}{Cp_m} \tag{54}$$

The working fluid temperature at dry-wet boundary line, T_{lb} [°C], cab be calculated from the energy balance equations between the air stream and working fluid in the dry region.

$$T_{lb} = T_{lo} - Y\,Cp_m(T_{ai} - T_{ab}) \tag{55}$$

Air temperature, air enthalpy, and working fluid temperature of dry region can be calculated by using traditional heat exchanger equations. But for the wet region, a discretized volume with distributed lump parameters method, like TFDC model that explained detail in Section 3, can calculate the parameters of air stream, saturated surface, and working fluid.

4.3.2 Energy balance between two fluids of Type-2.1.1 exchanger (energy balance line, EBL)

Energy balance line for the wet region of Type-2.1.1 exchanger can be derived by equating the energy lost from the air stream and the energy gained by the working fluid. EBL equation for each element with the numbering, n, of each node can be expressed as follows:

$$m_a dh_a = -m_l\,Cp_l dT_l \tag{56}$$

$$\frac{h_a^{n+1} - h_a^n}{T_l^{n+1} - T_l^n} = \pm\frac{m_l Cp_l}{m_a} = EBL \tag{57}$$

A minus sign refers to a parallel flow between air stream and working fluid flow. A plus sign refers to a counter flow that working fluid is flowing in the opposite direction of airflow.

4.3.3 Heat and mass transfer between two fluids of Type-2.1.1 model (tie-line slope)

Tie-line slope is the ratio of enthalpy difference between air and saturated surface to the temperature difference between working fluid and saturated surface. Tie-line slope, Eq. (59) can be derived from the equating of the sensible heat, transferring from the working fluid to saturated surface, with the total sensible and latent heat that transferring from the saturated surface to the air stream, see in

Figure 10. Tie-line slope of Type-2.1.1 extended surface TDCONF exchanger is similar with the one of TFDC exchanger, Eq. (19), but the difference is an addition of thermal resistances of the metal wall and fins in the wet region, Rmw. Hence, Tie-line slope for Type-2.1.1 extended surface TDCONF exchanger can be expressed with the characteristic of coil, *C*, as follows:

$$K_M(h_s - h_{aavg})dA = U_{sl}(T_{lavg} - T_s)dA \tag{58}$$

$$\frac{h_s - h_{aavg}}{T_s - T_{lavg}} = -\frac{U_{sl}}{K_M} = -\frac{Cp_m R_{aw}}{R_{mw} + R_l} = -\frac{1}{C} = -TLS \tag{59}$$

The total heat which is transferring from the saturated surface node j, T_s^j to the average temperature of air stream, T_{lavg}^j which is the average of two adjacent nodes temperature T_l^n and similarly from the saturated surface to the air stream are shown in **Figure 11**. *TLS* equation, Eq. (59), has two unknown parameters, saturated temperature, T_s, and saturated enthalpy, h_s. The third order polynomial regression equation, which is fitted with the saturation 100% RH line, shown in **Figure 7**, can be used as their relation equation Eq. (20) to the *TLS* equation Eq. (59), and rearranged with node number as follows:

$$coe1\left(T_s^j\right)^3 + coe2\left(T_s^j\right)^2 + (coe3 + TLS)\,T_s^j + coe4 - \left(\frac{h_a^n + h_a^{n+1}}{2}\right) - TLS\left(\frac{T_l^n + T_l^{n+1}}{2}\right) = 0 \tag{60}$$

This third order polynomial *TLS* equation can be solved by using Newton Raphson iteration methods to finds its unknown root parameter, T_s, from the known parameters of tie line slope, enthalpy of air, and working fluid temperature.

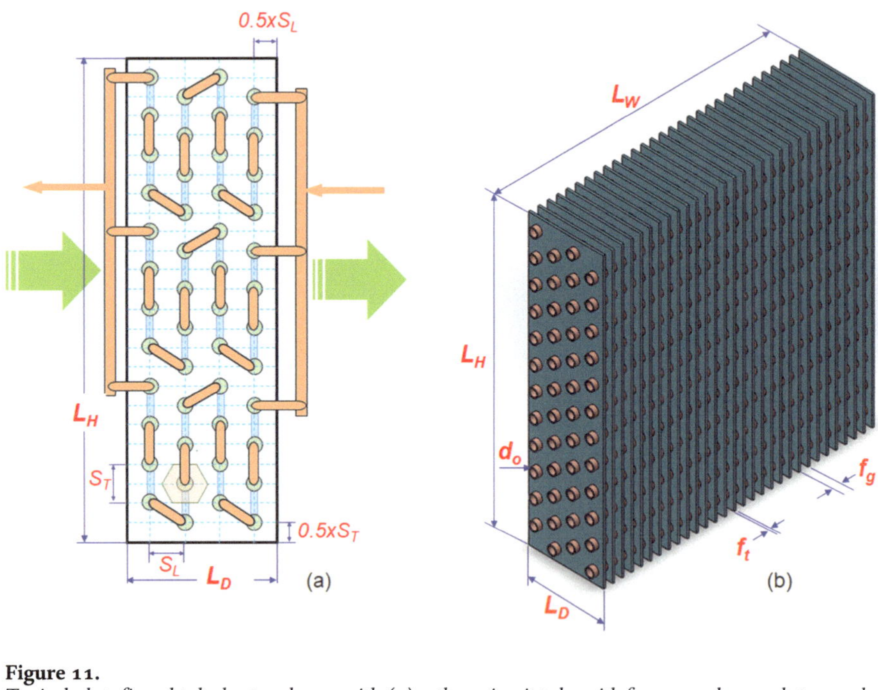

Figure 11.
Typical plate finned tube heat exchange with (a) a three circuit tube with four rows deep and staggered arrangement and, (b) its 3D view with dimensions' notation.

4.3.4 Dry bulb temperature of each node for Type-2.1.1 model

Likewise the derivation of air temperature in Section 3.3, the air temperature of each node of TDCONF exchanger can be driven from the ratio of sensible heat that transfer from air to saturated surface to the total heat and mass transfer from the air to saturated surface. The ratio is as follows:

$$\frac{m_a C_{pm} dT_a}{m_a dh_a} = \frac{\alpha_{aw}\left(T_s - T_{aavg}\right)dA}{K_M\left(h_s - h_{aavg}\right)dA} \tag{61}$$

The value of $\frac{\alpha_{aw}}{K_M C_{pm}}$ can be assumed 1 for the Lewis number to the power of 2/3 is similar with 1, see the derivation in Section 3.2. For numerical calculation, the air temperature of each node, Eq. (61) is needed to be rearranged and rewritten with node number n and j as follows:

$$T_a^{n+1} - T_a^n = \frac{\frac{h_a^{n+1}\ h_a^n}{\left(h_s^j - h_{avg}^j\right)}\left(T_s^j - T_a^n\right)}{\left(1 + \frac{h_a^{n+1} - h_a^n}{2\left(h_s^j - h_{avg}^j\right)}\right)} \tag{62}$$

From the above equation outlet air temperature of each node T_a^{n+1} can be calculated from the known parameters of air inlet temperature, air enthalpy, and saturated air enthalpy of each node.

4.3.5 Exchanger area calculation of Type-2.1.1

The area of wet region, A_w, can be calculated by integration of the energy balance equation which is the total heat losing from the air stream is equal to the total heat and mass transferring from the air stream to the vapor saturated surface layer.

$$\int m_a dh_a = \int K_M\left(h_s - h_{aavg}\right)dA \tag{63}$$

Mass transfer coefficient, K_M, can be substituted with the parameters $\frac{\alpha_{aw}}{Cp_m}$, as the Lewis number to the power of 2/3 is similar with 1, see in Eq. (6). The integral equation Eq. (63) can be solved out with Sampson's rule or Trapezoidal method, and it can rewrite with node number *n* and *j* as follows:

$$A_{ow} = \frac{m_a\ Cp_m}{\alpha_{aw}} \sum_{j=1}^{j=N-1}\left[\frac{1}{\left(h_s^j - h_{aavg}^j\right)} \times \left(h_a^{n+1} - h_a^n\right)\right] \tag{64}$$

The wet surface area of TDCONF exchanger can be calculated from the known parameters, air enthalpy h_a and vapor saturated enthalpy h_s of each element.

4.4 Type-2.1.2 extended surface TDCONF exchanger operating with volatile working fluid

The exchanger's geometrical structure of Type-2.1.2 is the same with the one of Type-2.1.1, for both are using plate finned tube heat exchanger having a similar thermal resistance for heat and mass transferring from air to working fluid. But the difference between them is that Type-2.1.1 exchanger is working with non-volatile fluid so that its fluid temperature T_L is increasing along the area axis from T_{Lin} to T_{Lout}, shown in **Figure 10**, and Type-2.1.2 exchanger is operating with volatile refrigerant so that its working fluid temperature is constant along the coil depth, dT_L is 0. As a result, the parameter Y become infinity. For coil characteristic C for Type-2.1.2 is similar with Type-2.1.1, the margin point, h_{ab}, of dry and wet region of Type-2.1.2 exchanger can be calculated from the coil characteristic equation Eq. (65). At boundary point, water saturated surface temperature of dry and wet region is same as dew point temperature of inlet air, $T_s = T_{dpai}$. The enthalpy of air at boundary, h_{ab} can be expressed as follows:

$$h_{ab} = h_{sb} + \frac{(T_{dpai} - T_l)}{C} \tag{65}$$

Similarly with Type-2.1.1, temperature of air at dry-wet boundary point can be calculated from the equation, Eq. (65). As the working fluid temperature is constant in all coil depth, fluid temperature at boundary point T_{Lb} is same as with T_L. Energy balance line for Type-2.1.2 is vertical for its refrigerant temperature is constant along the coil depth, dT_L is 0. As a result, the refrigerant temperature of each node is constant, and the enthalpy of air outlet h_{ao} can be calculated from the energy balance equation between the total heat lost from the air stream and the vaporization heat of the refrigerant, and the equation is as follows:

$$m_a dh_a = -m_l h_{fg(T_r)} \tag{66}$$

$$h_{ao} = h_{ai} - \frac{m_l h_{fg(T_r)}}{m_a} \tag{67}$$

The equations of tie-line slope, dry bulb air temperature, and contact area for Type-2.1.2 are the same with Eqs. (60), (62), and (64).

4.5 Type-2.2: non-extended surface TDCONF heat exchanger

Indirect evaporative cooler and M-cycle dew point evaporative cooler, Type-2.2 exchangers, are passive coolers and very efficient cooler for dry and hot region. The two fluids, a thin water film and working air, are directly contacting each other, and they do not contact with the third fluid, product air, by separating with aluminum or plastic film to prevent the moisture transferring and make sure allowing only heat transfer.

The phenomena of heat and mass transfer and its mathematical model is the same with a wet region of Type-2.1.1 exchanger, except the calculation of separator area. Tube is used as separator in Type-2.1.1 exchanger and flat sheet is used in Type-2.2 exchanger. Thus, total thermal resistance, RT, of non-extended surface TDCONF exchanger can be written with convective heat transfer coefficient of working air, α_{aw}, and of product air, α_{ap}. k_{sep} is the thermal conductivity of the separator and t_{sep} is the separator thickness. The total resistance equation can be expressed as follows:

$$R_T = R_{aw} + R_{mw} + R_{ap} = \frac{1}{\alpha_{aw}} + \left(\frac{1}{\alpha_{aw}} \frac{Cp_m}{m''}\right) + \frac{t_{sep}}{k_{sep}} + \frac{1}{\alpha_{ap}} \tag{68}$$

The rest equations of energy balance line (EBL), tie-line slope (TLS), dry bulb air temperature T_a, and contact area for Type-2.2 exchanger are the same with Eqs. (57), (60), (62), and (64).

5. Effectiveness of heat and mass exchanger

The effectiveness of heat and mass exchanger are defined in many different ways in many studies. The effectiveness of cooling tower, and direct evaporative cooler based on the temperature is the ratio of the range, the change of water temperature between the inlet, T_{Lin}, and the outlet, T_{Lout}, to the sum of the range and the approach which is the difference between water outlet temperature and the inlet air wet-bulb temperature, T_{WBain} [31], expressed as follows:

$$\varepsilon_{CoolingTower} = \frac{Range}{Range + Approach} = \frac{T_{Lin} - T_{Lout}}{T_{Lin} - T_{WBain}} \tag{69}$$

Although Jaber and Webb [8] proposed a modified definition of cooling tower effectiveness with enthalpy, temperature based effectiveness, Eq. (69), is widely used in the cooling tower industry and this study also. Saturation effectiveness is a key factor in the determination of direct and indirect evaporative cooler performance. The saturation effectiveness of direct evaporative cooler is the ratio of the dry bulb temperature difference between inlet air, T_{DBain}, and outlet air T_{DBaout}, to the difference between inlet air dry bulb temperature and its wet bulb temperature, T_{WBain} [2], expressed as follows:

$$\varepsilon_{DirectEvapCooler} = \frac{T_{DBain} - T_{DBaout}}{T_{DBain} - T_{WBain}} \tag{70}$$

The saturation effectiveness of indirect evaporative cooler, wet-bulb depression efficiency, is the ratio of the dry bulb temperature difference between the product air inlet, T_{DBpain}, and outlet, $T_{DBpaout}$, to the difference between the product air inlet and inlet wet bulb temperature of working air, T_{WBwain} [2], expressed as follows:

$$\varepsilon_{IndirectEvapCooler} = \frac{T_{DBpain} - T_{DBpaout}}{T_{DBpain} - T_{WBwain}} \tag{71}$$

The effectiveness of air washer chamber and wet part of cooling coil are based on the enthalpy, and can be defined as the ratio of enthalpy difference between air inlet and out let to the difference between inlet air enthalpy and saturation enthalpy associated with working fluid inlet temperature [8, 32], expressed as follows:

$$\varepsilon_{Coil/Airwasher} = \frac{h_{ain} - h_{aout}}{h_{ain} - hs_{WFin}} \tag{72}$$

6. Results and discussion

This study has developed four types of numerical models for DCHME widely used in the air-conditioning industry. The first model explained in Section 3 is for Type-1 TFDC exchanger: air washer chamber, cooling tower, swamp cooler, or direct contact evaporative cooler, shown in **Figures 2**, **5**, and **6**. The second model derived in Section 5.2 is for Type-2.1.1 extended surface TDCONF exchanger working with non-volatile refrigerant (2.1.1. Extended Surface-Non Vol). An example of these exchanges is the wet region of plate finned tube heat exchanger used for cooling and dehumidification working with chilled water or ethylene/propylene glycol, shown in **Figures 3(a)**, **8(b)** and **(c)**, and **10**. The third model derived in Section 5.3 is for Type-2.1.2 extended surface TDCONF exchanger working with volatile refrigerant (2.1.2. Extended Surface-Vol). Examples of these exchangers are the wet region of DX-coil for cooling and dehumidification working with R134a, R410, etc., shown in **Figures 3(a)** and **8(b)** and **(c)**. Finally, the fourth model, derived detail in Section 6, is for Type-2.2 non-extended surface TDCONF exchanger (2.2. Non-Extended Surface), which are indirect evaporative cooler and M-cycle dew point evaporative cooler, shown in **Figures 3(b)**, **8(a)** and **(c)**, and **12**.

There are two types of problems with known (given) parameters and parameters to calculate for DCHME, which are listed in **Table 1**. The first type of problem, Case 1a, and 1b, is the case for designing DCHME to solve the length or area of the exchanger to achieve the desired fluid outlet temp or air outlet condition. The second type of problem, Case 2, is the case for the model predictive control system to predict the effectiveness and output parameter of a given DCHME under the load's variation and different operational parameters. Model 1. TFDC and 2.1.1. Extended Surface-Non Volatile working fluids have three problem cases, but model 2.1.2. Extended Surface-Vol has only two problems excluded Case 1b because there is no temperature difference between the inlet and outlet of the volatile working fluid. Similarly, model Type-2.2. Non-Extended Surface has no Case 1a because the product air operating as a working fluid is the only parameter needed to design and develop the control model of the exchanger. The models with different types of problem cases are shown in **Figure 13**.

Figure 14 shows the simulation result of temperature and enthalpy of the air, water, and saturated layer of each element of Type-1 TFDC exchanger, which are cooling tower or air washer chamber with Case 1 problem, calculation of exchanger

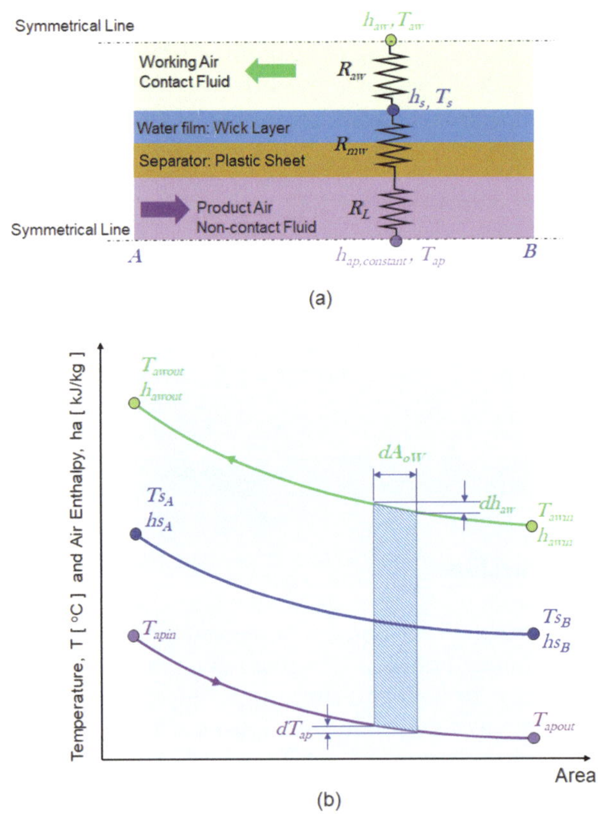

Figure 12.
(a) A typical one unit cell of non-extended TDCONF exchanger with counter flow process between product air and working air, and (b) thermal diagram of non-extended surface TDCONF exchanger and their potential difference for heat and mass transfer process.

DOI: http://dx.doi.org/10.5772/intechopen.102353

Problem types	Known parameters	Parameters to calculate
Case 1a [Parallel/Counter] *Calculation of length or area of exchanger to get the desired working fluid outlet temp.*	1. Both inlets [T_{ai}, w_i| T_{Li}] 2. Both flowrates [G_a, G_L] TFDC [m_a, m_L] TDCONF 3. Working fluid outlet [T_{Lo}] 4. Heat transfer coef. [α_a a_H, α_L a_H] TFDC [α_a, α_L] TDCONF	1. Air outlet condition [T_{ao}, w_o] 2. Saturate line [T_s, h_s] 3. Length [L] or area [A_w]
Case 1b [Parallel/Counter] *Calculation of length or area of exchanger to get the desired air outlet condition.*	1. Both inlets [T_{ai},w_i | T_{li}] 2. Both flowrates [G_a, G_l] 3. Air outlet condition [T_{ao}, w_b, h_a] 4. AW heat transfer coef. [$h_{aa}H$, $h_{la}H$] PFTHEX thermal resistance	1. Working fluid outlet [T_{lo}] 2. Saturate line [T_s, h_s] 3. Length [L] or area [A_w]
Case 2 [Parallel/Counter] *Prediction of the effectiveness and the working fluid outlet temp or air outlet condition of exchanger with known length or area.*	1. Both inlets [T_{ai}, w_i | T_{li}] 2. Both flowrates [G_a, G_l] 3. Length [L] 4. AW heat transfer coef. [$h_a a_H$, $h_l a_H$] PFTHEX thermal resistance	1. Both outlets [T_{ao}, w_o | T_{lo}] 2. Saturate line [T_s, h_s]

Table 1.
List of problem types with known and calculation parameters for DCHME.

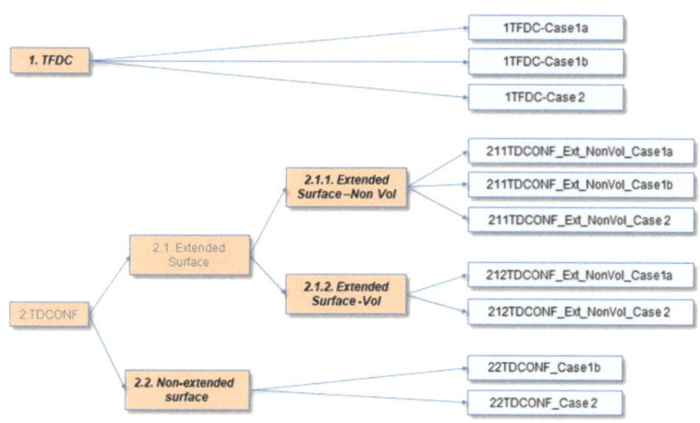

Figure 13.
Four models with three different problem types of air conditioning exchanger.

length or area. The numbering of nodes i and j for exchanger is shown in **Figure 5**, in which water is spraying into an (a) parallel or counter flow with airflow direction. The process is heating and humidifying airflow in the air washer chamber or cooling the water in the cooling tower. **Figure 14(a)** shows the result of parallel flow exchanger where the energy balance line is inclined to the left side due to its negative slope per Eq. (10), and similarly, the positive slope for counter flow, thus the line inclined to the right side as shown in **Figure 14(b)**. Both counter and parallel flow have the same negative tie line slope value, see Eq. (19), both tie lines are inclined into the left side. The outlet air temperature, green circle, of counter flow is 1.5°C higher than parallel flow under all the same condition. Thus, the counter flow has higher efficiency than the parallel flow.

Figure 15 shows the simulation result of temperature and enthalpy of the air, water, and saturated layer of each element of Type-1 TFDC exchanger, air washer

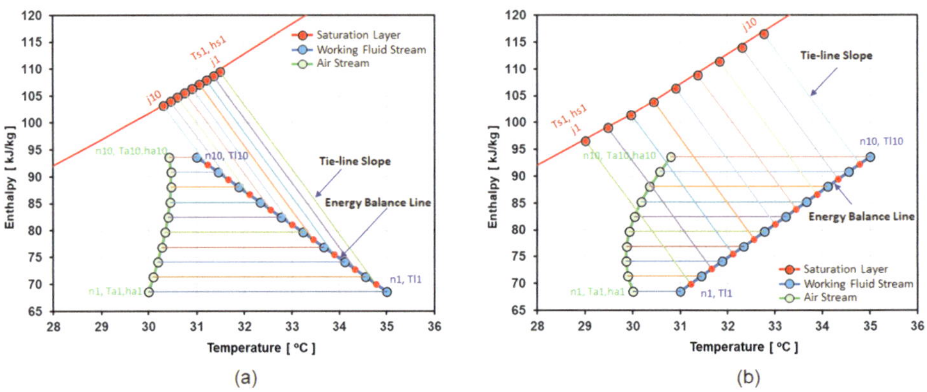

Figure 14.
Temperature and enthalpy of air, water and saturated layer of each element of Type-1 TFDC exchanger with Case 1 problem for the process of air heating and humidification in air washer chamber or cooling the water in cooling tower (a) parallel flow, and (b) counter flow.

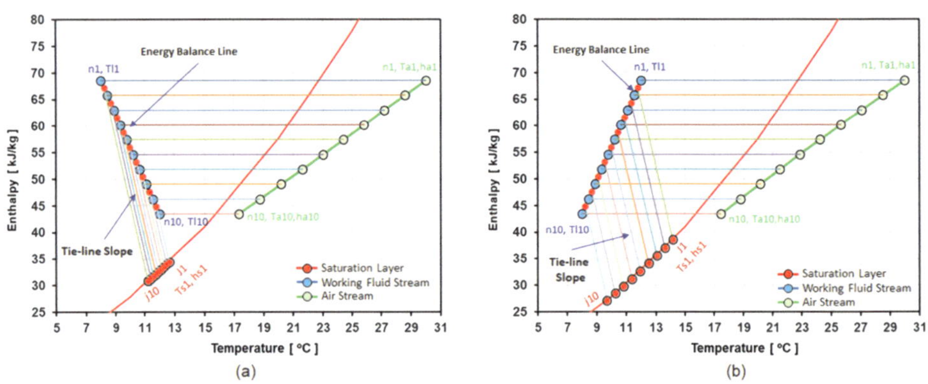

Figure 15.
Temperature and enthalpy of air, water and saturated layer of each element of Type-1 TFDC and Type-2.1.1 exchanger with Case 1 problem for the process of air cooling and dehumidification in air washer chamber and cooling coil operating with non-volatile working fluid in a (a) parallel flow, and (b) counter flow.

chamber, Type-2.1.1 exchanger, wet region of cooling coil working with non-volatile fluid, for the process of air cooling and dehumidification. The behavior of the energy balance line and tie line is the same as the heating and humidification process, but the waterline, blue circle, is left side of the saturation line because the working fluid temperature is lower than the saturation temperature, red circle, at each node. At the same time, the water line is on the right side of the saturation line because the working fluid temperature is higher than the saturation temperature in the air heating and humidification process, see **Figure 14**.

Figure 16 shows the effectiveness of the exchanger based on the water stream, Eq. (69), and based on the airflow stream Eq. (72) for cooling tower and air washer chamber under the same giving length with different liquid air ratio and inlet air enthalpy. At a low liquid-air flow ratio, the effectiveness of cooling tower based on water temperature stream is higher because the range, the temperature difference between water inlet and outlet, is higher due to the low water flow rate. However, for the air washer based on air stream, the effectiveness, Eq. (72), is lower than the higher liquid-air flow ratio because the temperature difference between the air inlet and outlet temperature is higher due to the high airflow rate. Therefore, this Type-1 exchanger with the Case 2 problem model can predict the effectiveness and outlet condition of water or air stream under the several of liquid-air flow ratio and inlet

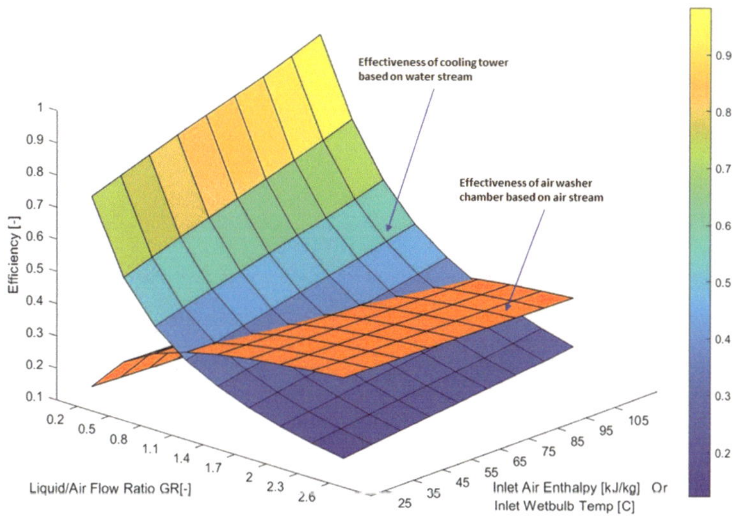

Figure 16.
Effectiveness of Type-1. TFDC exchanger based on the water stream for cooling tower and based on the air stream for air washer chamber: Type-1 exchanger with Case 2 problem model.

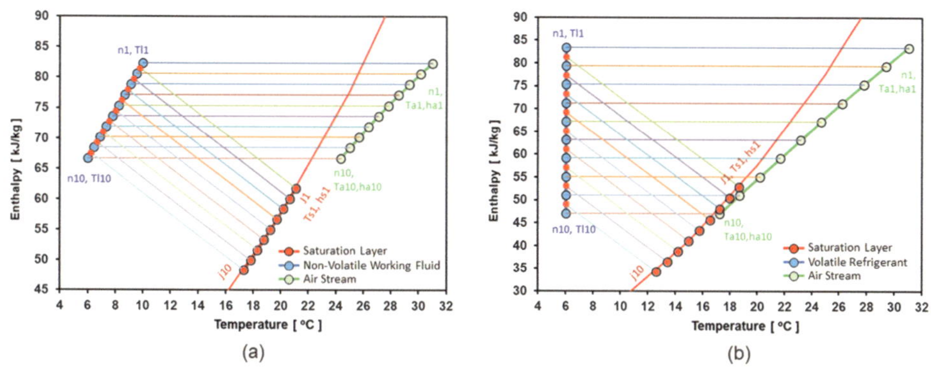

Figure 17.
Temperature and enthalpy of each element's (a) air, non-volatile working fluid and saturated layer of Type-2.1.1 exchanger (b) air, volatile working fluid and saturated layer of Type-2.1.2 exchanger with counter flow for the process of cooling and dehumidification.

air condition. Thus this model can be applied as a sub-function of the model predictive control system.

Figure 17 shows the simulation result of the temperature and enthalpy of each element. **Figure 17(a)** depicts the result of the model, Type-2.1.1 extended surface working with the non-volatile working fluid, and **Figure 17(b)** shows the result of the model, Type-2.1.2 extended surface working with volatile refrigerant for the process of cooling and dehumidification. Both models run at 6°C of working fluid temperature with the exact dimension of the plane finned tube heat exchanger and the same flow rate, flow rate ratio, and condition of the inlet air. Depending on the nature of the volatilization of the working fluid, the temperature of the non-volatile working fluid changes from inlet 6°C to outlet 10°C. The outlet air temperature is dropped from 31 to 24.1°C, as shown in **Figure 17(a)**, whereas, in the model of Type-2.1.2, the refrigerant temperature is constant at 6°C, but the outlet air temperature is dropped from 31 to 17°C, as shown in **Figure 17(b)**. Thus, the Type-2.1.2 exchanger has a better performance than the Type-2.1.1 exchanger.

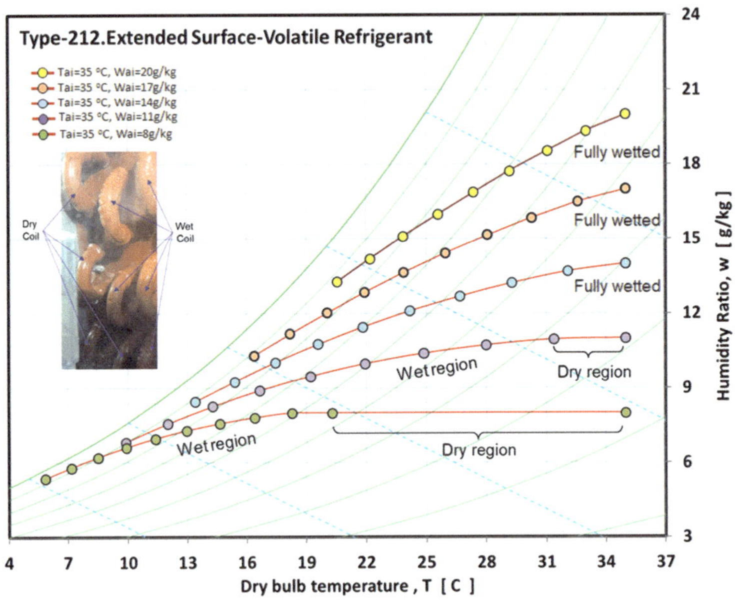

Figure 18.
Cooling and dehumidification process of Type-2.1.1 exchanger with same inlet air temperature but different humidity ratio.

Figure 18 shows the simulation result of model Type-2.1.1 exchanger running with the same plate finned tube heat exchanger and operating with the same volatile refrigerant at 5°C with the same refrigerant-air flow ratio. The result clearly shows that the model can estimate the dry and wet area and predict the outlet air temperature and moisture removing rate under different air inlet conditions (same inlet temperature but different humidity ratio). Thus, these models are suitable for use as a sub-function of the model predictive control system.

7. Conclusion

This study has developed a mathematical model based on a discretized volume with distributed lumped-parameters method for two fluid direct contact (TFDC) exchangers and two direct contacts with one non-contact fluid (TDCONF) exchanger. Based on the flow system and structure, this study has developed four models; Type-1 TFDC exchanger model (air washer chamber, cooling tower, and swamp cooler or direct contact evaporative cooler), Type-2.1.1 extended surface TDCONF exchanger model working with non-volatile refrigerant (wet region of plate finned tube heat exchanger cooling coil working with chilled water or ethylene/propylene glycol), Type-2.1.2 extended surface TDCONF exchanger working with volatile refrigerant (wet region of plate finned tube DX-coil), and Type-2.2 non-extended surface TDCONF exchanger (indirect evaporative cooler and M-cycle dew point evaporative cooler). From the simulation result, these models can reflect both heat and mass transfer behavior in every spatially distributed physical system. Moreover, they can predict well the effectiveness and dependent parameters of DCHME under the different load conditions and its various input parameters. Hence, these models can be a valuable tool for designing the exchangers mentioned above and can be applied as a sub-function of the model predictive control system.

Author details

Marip Kum Ja[1]*, Qian Chen[1], Muhammad Burhan[1], Doskhan Ybyraiymkul[1], Muhammad Wakil Shahzad[2], Raid Alrowais[3] and Kim Choon Ng[1]

1 Water Desalination and Reuse Center, King Abdullah University of Science and Technology, Thuwal, Saudi Arabia

2 Northumbria University, Newcastle Upon Tyne, United Kingdom

3 Civil Engineering Department, Al Jouf University, Skaka, Saudi Arabia

*Address all correspondence to: mkum.ja@kaust.edu.sa

References

[1] Jacobs HR. Direct contact heat exchangers. DOI:10.1615/AtoZ.d.direct_contact_heat_exchangers

[2] 2020 ASHRAE Handbook—HVAC Systems and Equipment (SI). Atlanta, GA: ASHRAE; 2020

[3] Lin J, Wang RZ, Kumja M, Bui TD, Chua KJ. Modelling and experimental investigation of the cross-flow dew point evaporative cooler with and without dehumidification. Applied Thermal Engineering. 2017;**121**:1-13

[4] Lin J, Thu K, Bui TD, Wang RZ, Ng KC, Kumja M, et al. Unsteady-state analysis of a counter-flow dew point evaporative cooling system. Energy. 2016;**113**:172-185

[5] Lin J, Wang RZ, Kumja M, Bui TD, Chua KJ. Multivariate scaling and dimensional analysis of the counter-flow dew point evaporative cooler. Energy Conversion and Management. 2017;**150**:172-187

[6] Kays WM, London AL. Compact Heat Exchangers. New York: McGraw-Hill Companies; 1964

[7] London AL, Mason WF, Boelter LMK. Performance characteristics of a mechanically induced draft, counter flow, packed cooling tower. Transactions of ASME. 1940;**62**:41-50

[8] Jaber H and Webb BL. Design of cooling towers by the effectiveness-NTU method. Journal of Heat Transfer. November 1989;**111**(4):837-843

[9] Berman LD. Chapter 2. In: Sawistowski H, editor. Evaporative Cooling of Circulating Water. 2nd ed. New York: Pergamon Press; 1961. pp. 94-99. [translated from Russian by R. Hardbottle]

[10] Webb RL. A critical review of cooling tower design methods. In: Shah RK, Subbarao EC, Mashelkar RA, editors. Heal Transfer Equipment Design. Washington, DC: Hemisphere Pub. Corp.; 1988. pp. 547-558

[11] Whillier A. A fresh look at the performance of cooling towers. ASHRAE Transactions. 1976;**82**:269-282

[12] Agarwal A, Lang J. Course Materials for 6.002 Circuits and Electronics. MIT OpenCourseWare (PDF). Boston: Massachusetts Institute of Technology; 2007

[13] Groppia G, Bellolie A, Tronconip E, Forzatti P. A comparison of lumped and distributed models of monolith catalytic combustors. Chemical Engineering Science. 1995;**50**(17):2705-2715

[14] Ja MK, Koyama S. Numerical and experimental investigations on a two-bed mode adsorption chiller, ICR 2007. Refrigeration creates the future. In: Proceedings of the 22nd IIR International Congress of Refrigeration; 21–26 August 2007; Beijing, China

[15] Ja MK, Choo FH, Li B, Chakraborty A, Dass ETM, Zhao K, et al. Two-dimensional numerical analysis of membrane-based heat and mass cross-flow exchanger. Heat Transfer Engineering. 2017;**38**(4):438-445

[16] Kumja M, Ng KC, Yap C, Yanagi H, Koyama S, Saha BB, et al. Modeling of a novel desorption cycle by dielectric heating. Modern Physics Letters B. 2011; **23**(3):425-428

[17] Kum Ja M. Performance of microwave-activated adsorption cycle: Theory and experiments [Ph.D. thesis]. 2010. Available from: http://scholarbank.nus.edu.sg/handle/10635/18640

[18] Dimitra Psychogiou, Roberto Gómez-García, Raúl Loeches-Sánchez, Dimitrios Peroulis. Hybrid

acoustic-wave-lumped-element resonators (AWLRs) for high-Q bandpass filters with quasi-elliptic frequency response. IEEE Transactions on Microwave Theory and Techniques. JULY 2015; **63**(7):447-449

[19] 2001 ASHRAE Fundamentals Handbook (SI), Section 5.9. Simultaneous Heat and Mass Transfer Between Water-Wetted Surfaces and Air. Atlanta, GA: ASHRAE; 2001

[20] ASHRAE Systems and Equipment Handbook (SI). Atlanta, GA: ASHRAE; 2016

[21] Bird RB, Stewart WE, Lightfoot EN. Transport Phenomena. New York: John Wiley and Sons; 1960

[22] Incropera FP, DeWitt DP. Fundamentals of Heat and Mass Transfer. 4th ed. New York: John Wiley and Sons; 1996

[23] Chilton TH, Colburn AP. Mass transfer (absorption) coefficients—Prediction from data on heat transfer and fluid friction. Industrial and Engineering Chemistry. 1934;**26**:1183

[24] Bedingfield GH Jr, Drew TB. Analogy between heat transfer and mass transfer—A psychometric study. Industrial and Engineering Chemistry. 1950;**42**:1164

[25] McElgin J and Wiley DC. Calculation of coil surface areas for air cooling and dehumidification. Heating, Piping, and Air Conditioning. March 1940;**12**(3):195

[26] Kusuda T. Graphical method simplifies determination of air coil, wet heat-transfer surface temperature. Refrigerating Engineering. 1957;**65**:41

[27] 2001 Standard for Forced-Circulation Air-Cooling and Air-Heating Coils, AHRI Standard 410-2001. Arlington, VA: Air-Conditioning, Heating and Refrigeration Institute; 2001

[28] Schmidt TE. La Production Calorifique des Surfaces Munies d'ailettes, Annexe, Du bulletin De L'Institut International Du Froid, Annexe G-5. 1945

[29] Brown G. Theory of Moist Air Heat Exchangers. vol. 77. Stockholm, Sweden: Transactions of the Royal Institute of Technology; 1945. pp. 12-15

[30] Ware CD, Hacha TH. Heat Transfer From Humid Air to Fin and Tube Extended Surface Cooling Coils, ASME Paper No. 60-HT-17. New York, NY, USA: American Society of Mechanical Engineers; 1960

[31] Cheremisinoff NP, Cheremisinoff PN. Cooling Towers: Selection, Design, and Practice. Ann Arbor, MI: Ann Arbor Science Publishers; 1981

[32] Nellis G, Klein S. Heat Transfer. Cambridge: Cambridge University Press; 2008

Section 4

Optimization of Geometry

Chapter 6

Assessment of Augmentation Techniques to Intensify Heat Transmission Power

Prashant B. Dehankar

Abstract

The heat exchanger detects heat between two processes of liquids in the chemical, petrochemical, food, beverage, and hot metals, and so on. Although the required heat transfer calculations and pressure reductions are achieved with a two-pipeline temperature switch (DPHE), the optimization of the heat transfer parameter is used to measure laboratory test settings. This will allow you to build a DPHE model with twisted tapes and mimic the ASPEN PLUS and work it out by trying to scale the lab that has already been produced and standardized for DPHE. Parameter values for this study range from 0.02 kg / sec −0.033 kg / sec as suspension, pressure reduction, and Reynolds numbers. Also to study the mechanism of increased heat transfer by the use of twisted tape with Y1 = 4.3 and Y2 = 7.7 deviations. They are trying to compare the results of a mathematical model with simulation. This mode of inactivity has the effect of equilibrium heat transfer, pressure drop, the collision factor, and the number Reynolds. We tested the modeling and simulation effects and tried to measure the 4 input parameters of the two output parameters: cold flow rate, hot flow rate, cold and cold temperatures. DPHE, therefore, confirmed the flow rates of weight between 0.02–0.07 kg/s with experiments and simulations performed by Aspen Plus.

Keywords: modeling, simulation, DPHE, twist tape, ASPEN PLUS

1. Introduction

Heat exchangers have many applications in industry and engineering. The process of building a heat exchanger is somewhat difficult because, apart from problems such as long-term operation and the economic side of the equipment, careful study of the rate of heat transfer and pressure drop pressure is required. The biggest problem in building a heat exchanger is to make the equipment compact [1], with less pumping power reaching a higher transmission rate. In a variety of technological applications, **Figure 1**, heat transfer techniques are compatible. The high cost of energy and materials in recent years has led to a concerted effort to produce more efficient heat exchange equipment. In addition, for special applications such as space use, a reduction in temperature is sometimes required by increasing the heat transfer. While changes in fluid flow (viscosity breakage and thermal boundary layers) can increase the heat transfer rate, the pumping capacity in this process can increase significantly and ultimately the cost of pumping. Therefore, many

Figure 1.
The idea of a model.

techniques have been developed to obtain the required amount of heat transfer to an existing heat exchanger with low-cost pumping power. A process model is a set of statistics that allows us to predict the performance of a chemical process system.

Increasing heat transfer, which leads to energy and cost savings, is of paramount importance to academics. In the field of processes and engineering, there are many devices used to transfer heat to stations. With many additional strategies, conventional exchanges are developed with an emphasis on various types of site improvement [2, 3]. Extra fixtures can help to increase the heat transfer rate and the unwanted increase in conflicts with one or more of the following:

- Disruption and disruption of energy levels.
- The heat transfer area is increased.
- Generate rotating/rotating/secondary flow.

2. Objectives

1. Developing a DPHE model.
2. To conduct a DPHE test to validate the model.
3. The use of the results comparison software.

Enhanced surfaces can generally be used for three purposes,

a. The pumping power required for the heat transfer process should be reduced.

b. Increase the heat exchanger's total UA value.

c. Creating model equations for hot and cold fluid output temperature.

In any case, a greater UA value can be used,

- To achieve a greater heat change at the fixed temperature of the fluid inflow, or
- Reduces the mean heat exchange temperature differential by increasing thermodynamic process efficiency that might lead to operational costs being saved.

3. Motivation

For many years efforts have been made to create more heat exchangers using various methods to improve heat transfer. Due to the increasing demand for the industry in terms of heat exchange devices, however, they are lower in design and

Figure 2.
Types of augmentation techniques.

operation than conventional heat exchange devices; advanced heat transfer studies **Figure 2** have become more effective in recent years. Energy-saving and energy consumption also provide a significant incentive for new development processes. It is important that heat exchangers are cohesive and lightweight when building cooling systems for cars and spacecraft. Additional devices are also required for high-temperature power switches (i.e. air-cooled condensers, nuclear fuel rods). This and many other uses have led to the production of various improved heat exchangers.

More heat transfer is greatly increased by retransmission or redistribution of flow to improve efficiency,

- Axial Reynolds number
- Shortcut flow area
- It means speed
- Temperature gradient near the wall of the tube

4. Methodology

Existing enhancement techniques can be broadly classified into three different categories:

1. Active Techniques: some external power to improve the heat transfer rate in the heat exchanger is appropriate for that process. This method is no more complicated than the synthetic and complex method, for example, machinery, surface vibration, liquid vibration, etc.

2. Passive Techniques: The flux pattern changes in this method only with systemic power available without external power. This change causes the layer of temperature disturbances and pressure to decrease to improve the heat transfer rate of the heat exchangers e.g. heavy face, swirl flow, etc.

3. Compound Techniques: An effective and practical process is used in this approach. Complex technology is a complex that raises the pressure and decreases, for example, an empty tube with twisted tape, a water tube, and so on.

5. Observations from literature review

Twisted tape installation mixes the flow of quantities well and is thus superior to any other installation in laminar flux, as heat resistance is not restricted from laminar flow [4] to a small area near the wall. But the performance of twisted tape

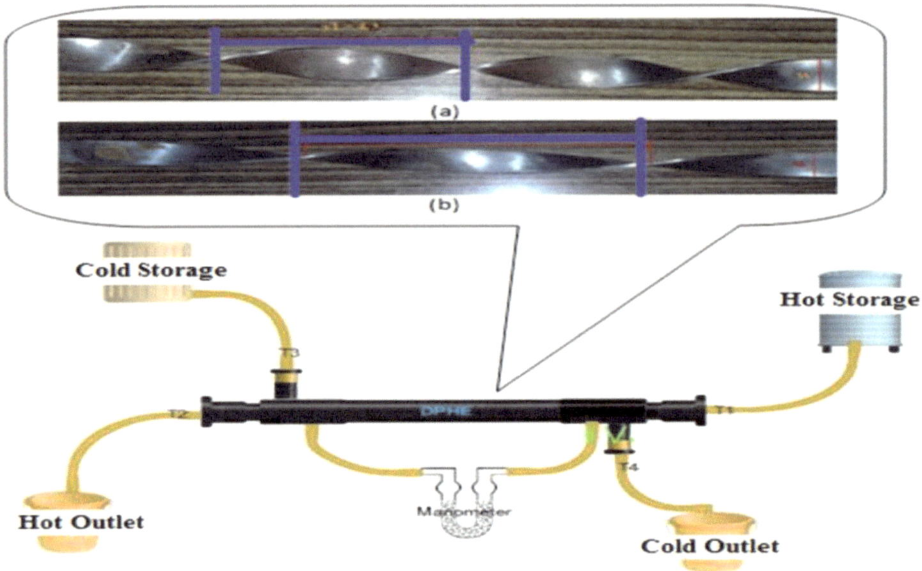

Figure 3.
Experimental DPHE setup with twisted tape.

depends on the fluid parameters such as Prandtl number. Due to its flexible volume, the length of a twisted band **Figure 3(a)** and **(b)** is better than the length of a full band. Twisted tape can be used effectively to increase heat transfer when designing small heat exchanges **Figure 3** for laminar flow.

Turbulent flow [5] twist tape works well up to a certain level of Reynolds but not beyond the wide range of Reynolds. The twisted tape does not work in a turbulent flow compared to a wind turbine due to the drop in pressure drop. So the performance of hot rolled tape is not very good compared to the turbulent flow wire coil. It can therefore be proved that in a turbulent environment a wire spool is a good choice. However, short twisted tapestry offers better hydrothermal performance compared to long and twisted tapestry.

6. Selection and choice of process

The selected process consists of a liquid flow, e.g. water through a double body temperature. The heat exchanger has a lot of industrial and engineering performance. An accurate study of heat transfer rate and pressure reduction and performance and economic aspect of heat exchange equipment is required in the development of a heat exchange process [6–8]. The pressure drop increases when the input is used to increase the heat transfer and the heat transfer rate. As the pressure decreases, pumping costs are high [9–12]. It is also very important that the pressure drop does not exceed a certain amount when using the insert to install a heat transfer system.

7. Mathematical modeling

7.1 Hypothesis

To expand the mathematical model, we look at this simple idea to do the following:

- Merchant performance in a consistent state government.
- Transmission of heat to the environment is ignored.
- HE is considered to be a system with illuminated parameters.
- Two flows in the liquid phase and does not change the phase.

7.2 Modeling equations

By taking heat balance of hot & cold fluid, we get,

$$Q_h Cp_h\,(T_1 - T_2) = Q_c Cp_c\,(T_4 - T_3) \tag{1}$$

$$Q_h Cp_h\,(T_1 - T_2) = U\,A\Delta T_{lm} \tag{2}$$

$$Q_h Cp_h\,(T_1 - T_2) = \frac{U\,A\,\{\,(T_1 - T_4) - (T_2 - T_3)\}}{\ln\,\frac{(T_1 - T_4)}{(T_2 - T_3)}} \tag{3}$$

The heat exchanger's mathematical model has been constructed, and it includes a heat balance equation for the two material fluxes Qh and Qc, as well as an expression for transferred heat flow **Table 1**. The overall heat exchange coefficient, U, has a standard expression as the overall HTC, which may be given by Eq. (4), for the heat flow transported in the heat exchanger.

$$U = \frac{1}{\left\{\left(\frac{1}{h_i}\right)\left(\frac{d_e}{d_i}\right) + \left(\frac{d_e}{2k}\right)\ln\left(\frac{d_e}{d_i}\right) + \left(\frac{1}{h_o}\right)\right\}} \tag{4}$$

7.3 Solving of the mathematical model

Eq. (1) represents a system of two non-linear equations with two variables having the form,

$$f_1\,(T_2, T_4) = 0,\;\; f_2\,(T_2, T_4) = 0 \tag{5}$$

$$f_1 = Q_h Cp_h\,(T_1 - T_2)\;\; - Q_c Cp_c\,(T_4 - T_3) \tag{6}$$

$$f_2 = \{Q_h Cp_h\,(T_1 - T_2)\}\;\; - \left\{\frac{U\,A\,\{\,(T_1 - T_4) - (T_2 - T_3)\,\}}{\ln\,\frac{(T_1 - T_4)}{(T_2 - T_3)}}\right\} \tag{7}$$

The equation's unknown variables (4), the hot fluid's T2 outlet temperature, and the cold fluid's T4 outlet temperature, are also the heat exchanger's output variables. The expressions of the functions f1 and f2 of the Eq. (5) are defined by the relations (6) and (7).

T1° C	T2° C	T3° C	T4° C	Nre	U W/m^2 K
70	49.25	29	38	11936.60	169.37
60	47.48	29	37	13433.33	171.65
55	45.1	29	35	13602.49	172.51

Table 1.
Experimental parameters.

We can get solutions from two methods,

- Jacobean matrix
- Newton–Raphson method

The Jacobean matrix associated with the linear Eq. (5) is given by,

$$J(X) = \begin{bmatrix} \frac{\partial f_1}{\partial T_2} & \frac{\partial f_1}{\partial T_4} \\ \frac{\partial f_2}{\partial T_2} & \frac{\partial f_2}{\partial T_4} \end{bmatrix} \tag{8}$$

$$\frac{\partial f_1}{\partial T_2} = -Q_h Cp_h \tag{9}$$

$$\frac{\partial f_1}{\partial T_4} = -Q_c Cp_c \tag{10}$$

$$\frac{\partial f_2}{\partial T_2} = -Q_h Cp_h - \left\{ \frac{[U\,A\,\{\,(T_1 - T_4) - (T_4 - T_3)\,\}]\left[\frac{(T_1-T_4)}{(T_2-T_3)}\right]^2}{\ln\left[\frac{(T_1-T_4)}{(T_2-T_3)}\right]^2} \right\} \tag{11}$$

7.4 Example a certain amount when using the insert to install a heat transfer system

Assuming, Temperature T1 = 70°C and T3 = 29°C; μ at 70̊ C = 0.0004101 Pa ses; k = 0.62136 W/m K; de = 0.0280 m; di = 0.013 m;
Cp = 4186 J/kg K.
Apply Eq. (5)

$$h_i = \frac{j^H\ k}{d_i\ \left(C_p\ \frac{\mu}{k}\right)^{1/3}}$$

$$h_o = \frac{j^H\ k}{d_e\ \left(C_p\ \frac{\mu}{k}\right)^{1/3}}$$

$$N_{re} = \frac{d_i\ v\ \rho}{\mu}$$

$$v = \frac{Q}{A}$$

By using Eqs. (1), (2) and (3) and substituting all above calculated values we get,

$$\dot{m}\ C_p\ \Delta T_{|hot} = \dot{m}\ C_p\ \Delta T_{|cold}$$

T2 = 190.35–4.15 T4

$$\dot{m}\ C_p\ (T_1 - T_2) = U\ A\Delta T_{lm}$$

By substituting value of T2 from Eq. (1)

$$f(T_4) - \frac{(70 - T_4) - (161.35 - 4.15\ T_4)}{\ln(70 - T_4) - \ln(161.35 - 4.15\ T_4)} - 66.15\ T_4 + 1918.07$$

From trial and error method,
T4 = 38° C.
From Eq. (1), substituting T4.
T2 = 49.25° C.

8. Results of mathematical model

8.1 About the inserts

Mild steel twisted tapes were utilized as inserts in the experiment **Table 2**. The purpose of this study is to determine the friction factor and heat transfer coefficient for twisted tape with a twist ratio of (Y1 = 4.3, Y2 = 7.7) and compare them to those of a smooth tube.

Twist ratio 1 = 4.3.
Twist ratio 2 = 7.7.
Twisted tape thickness = 0.002 m.
Twisted tape length = 0.90 m.
Twist width = 0.012 m.

8.2 Simulation

Simulation means the equation through the use of software mathematical tools. Chemical engineering requires a process simulation to solve difficulties associated with process design, process analysis control, and much more, in the actual world a chemical process is defined in a process fluid sheet.

What is Process Simulation/Analysis?

The analysis/simulation is designed to model and predict how the process is performed. In any performance test, the process must be broken down into fragments (e.g. units). Process factors are predicted by analytical strategies (e.g. flow rates, tracks, temperatures, pressures, features, equipment size, etc.) Mathematical models, dynamic correlations, and process simulation tools include these strategies (e.g. ASPEN Plus). In order to predict and validate performance, process analysis may include the use of test methods. As a result, we obtain the installation process and the flow process during the process simulation and are required to predict the results of the process **Figure 4**. ASPEN Plus is the focus of the lab. It is computer-assisted software that predicts process performance (e.g., broadcast parameters, operating conditions, and machine sizes, and uses basic

Inner pipe ID	0.013 m
Inner pipe OD	0.015 m
Outer pipe ID	0.023 m
Outer pipe OD	0.025 m
MOC of tube	Cu
MOC outer pipe	PVC
Heat transfer length	0.90 m
Outer pipe length	0.76 m

Table 2.
Specification of experiment setup.

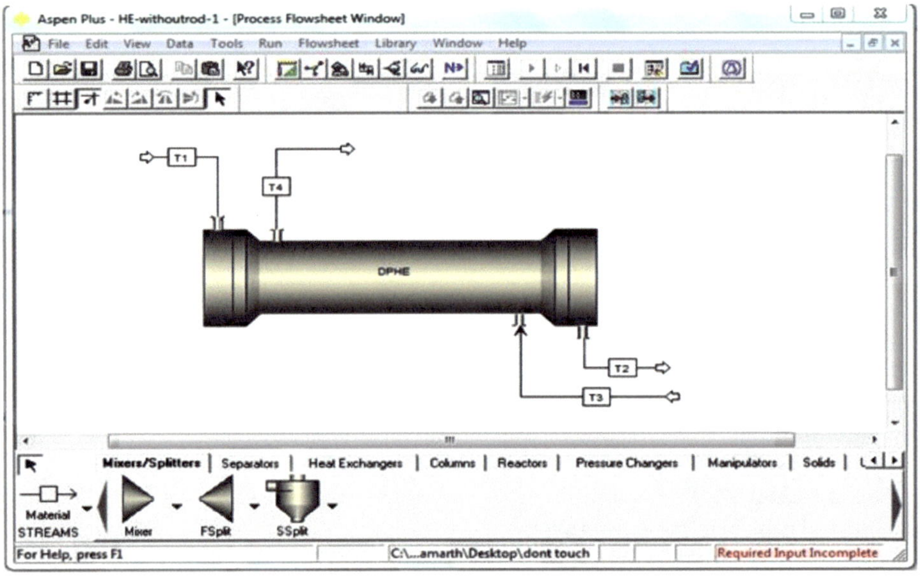

Figure 4.
DPHE with inlet and outlet streams.

physical coordinates (e.g., material and power balances, thermodynamic equilibrium, and measurement).

Computer-assisted simulation has various benefits:

- Allows designers to easily evaluate and provide information on integrated water systems integration.

- It can be integrated to achieve a complete integrated design and integration process.

- Reduces testing and expansion efforts.

- Flexibility and awareness of the process are assessed by answering the questions of "what if?"

- Modeling process and understanding process performance in bulk.

9. Result and discussion

9.1 Final result

At DPHE, the set results were confirmed by experimental work on mathematical models **Tables 3** and **4**. **Figures 5** and **6** shows that, in comparison with the test values for the full total result, a larger 4°C rating was obtained for the mathematical model due to a few manual errors.

Mathematical model analysis and test function are shown in **Figures 7** and **8**. The test consists of a smooth tube with a torsioned tube Y1 = 4.3 and Y2 = 7.7 without installation and installation. Recovery of heat from hot and cold liquids increases as **Table 5** temperature differences increase. Extreme heat transfer from

T2° C	T4° C	U W/m² K
49.25	38	169.37
47.48	37	171.65
45.1	35	172.51

Table 3.
Theoretical values for temperatures and heat transfer coefficient.

Smooth tube			$Y_1 = 4.3$			$Y_2 = 7.7$		
T2	T4	U	T2	T4	U	T2	T4	U
55	35	1849.002	56	39	2941.53	55	39	2456.17
51	34	2505.59	48	34	3762.55	50	33	2859.64

Table 4.
Experimental values of heat transfer coefficient for smooth and twist tape.

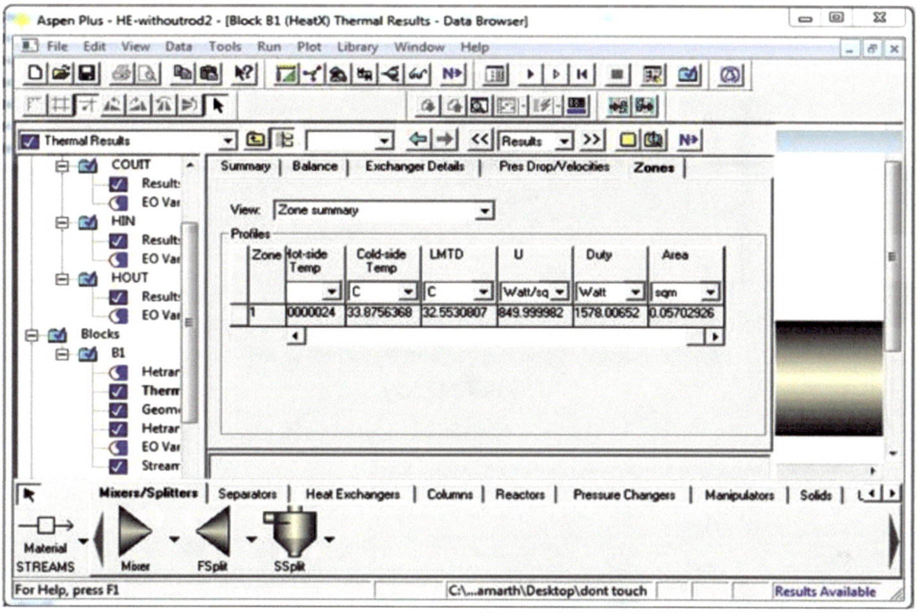

Figure 5.
Thermal results in plane tube simulation.

hot and cold liquids is used in the form of heat transfer. It was noted that the 6°C of the twist ration Y1 = 4.3 and the 4°C of the twist ratio Y2 = 7.7 as the comparable smooth tube was increased using twist-tape.

Figure 9 translates that, the conflict decreased as the snumber of Reynolds increased. Therefore, when twisted tapes with different twist ratios are shown in 4.3 and 7.7 a device that is more important than a smooth tab is used. The heat transfer coefficient with a twist ratio of 4.3 rather than a 7.7 twist ratio. Therefore, the ratio of the contraction of the pulsating tap and the sugar is very smooth to 4.3 twist ratio and then 7.7.

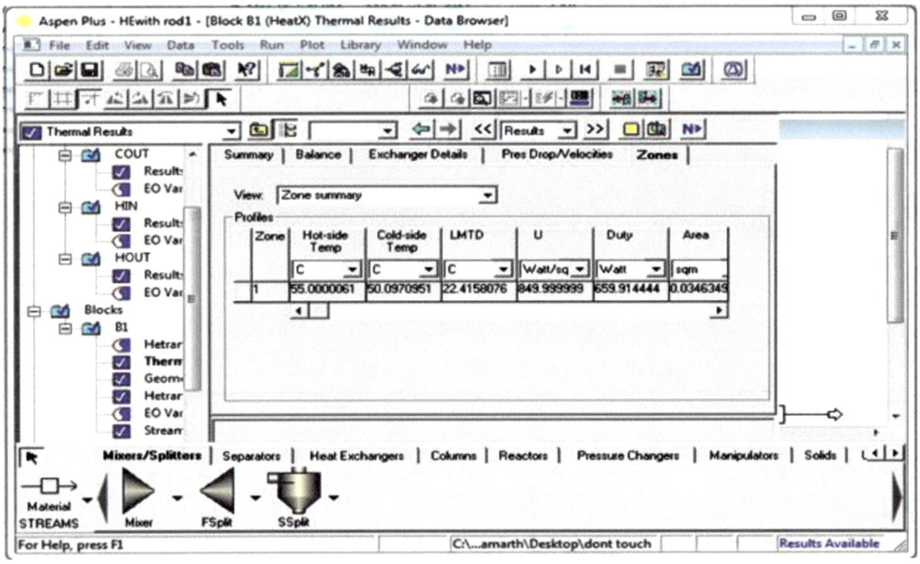

Figure 6.
Thermal results in tube with tape simulation.

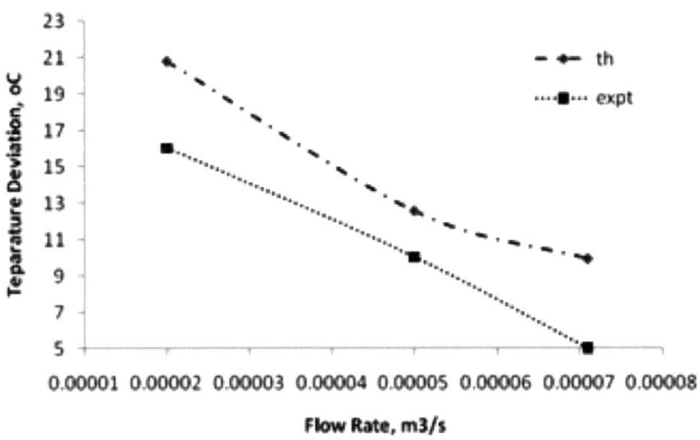

Figure 7.
Comparison between mathematical model and experimental work.

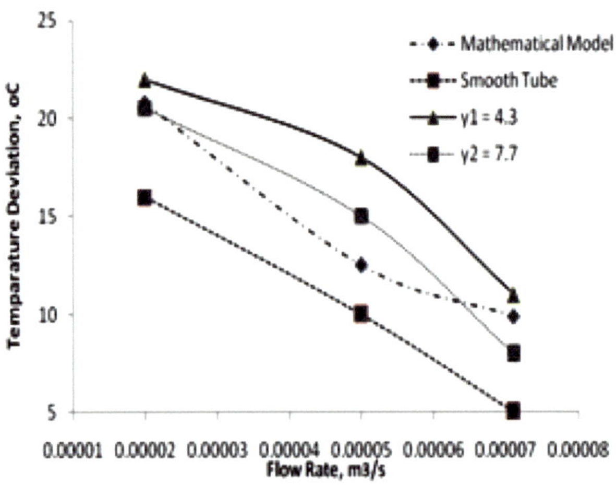

Figure 8.
Comparison between mathematical model and experimental work for smooth tube, twisted tape having twist ration (Y1 = 4.3 & Y2 = 7.7).

Flow rate		Model		Smooth tube			$Y_1 = 4.3$			$Y_2 = 7.7$		
	T1	T2	ΔT	T1	T2	Δt	T1	T2	ΔT	T1	T2	ΔT
0.00002	70	49.25	20.75	74	58	16	71	49	22	70	49.5	20.5
0.00005	60	47.48	12.52	65	55	10	63	45	18	62	47	15
0.000071	55	45.1	9.9	56	51	5	59	48	11	60	52	8

Table 5.
Comparison of temperature difference.

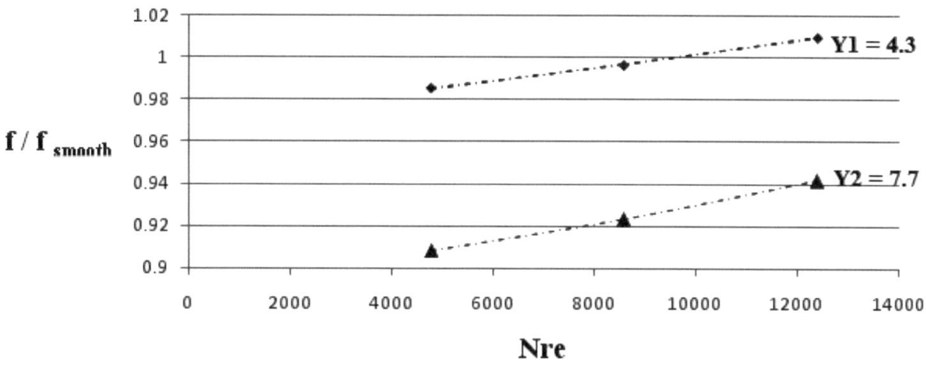

Figure 9.
Friction factor ratios vs. Nre for theoretical and experimental values.

10. Conclusion

In this experiment, we find that with the introduction of a torn tape, the heat transfer rate increases. When the installation is installed in a flow-through station, the rate of heat transfer and pressure drop leads to a higher level of turbulence. By comparing experiments using open source tools, we get results similar to Aspen Plus.

The statistical model is validated by comparing the test values under the same operating conditions of the two output flow temperatures associated with the model. In the analysis of industrial vendors in the cleaning areas, a mathematical model and algorithm of its solution will be used.

Scope for Future Work

1. DPHE manufactured to compute temperature-related viscosity; for energy saving.

2. As a process of intensification approaching the green technology principle in the chemical industry.

Nomenclature

Ai	Inside heat transfer surface area, m^2
Ac	Cross-sectional area, m^2
Cp	Specific heat of fluid, J/kg K
Di	Inside diameter of the tube, m

fa	Fanning friction factor, dimensionless
f	Theoretical Fanning friction for smooth tube, dimensionless
Gz	Greatz number, dimensionless
h	Difference in level of CCl4 in the manometer, m
hi	Inside HTC, W/m^2 K
ho	Outside HTC, W/m^2 K
H	Linear distance of the tape for 180° rotation
k	Thermal conductivity of the fluid, W/m K
Lw	Heat transfer length, m
Lh	Pressure taping to pressure taping length, m
ΔTlm	Log mean temperature difference
$\dot{m}$	Mass flow rate, kg/s
Nu	Nusselt number
Pr	Prandtl number
ΔP	Pressure drop
Q	Heat transfer rate, W
Re	Reynolds number
T	Temperature in° C
Ui	Overall heat transfer coefficient, W/m2 K
v	Velocity of water, m/s
w	Width of the twisted tape
Wt	Weight of the water taken, kg
y	Twist ratio, dimensionless, H/d

Greek letters

ρ	Density of fluid, kg/m^3
μ	Dynamic viscosity of the fluid, Pa sec

Author details

Prashant B. Dehankar
Tatyasaheb Kore Institute of Engineering and Technology, Warananagar, Shivaji University, Kolhapur, Maharashtra, India

*Address all correspondence to: dehankarpr@gmail.com

References

[1] Stone KM. Review of literature on heat transfer enhancement in compact heat exchangers. ACRC. 1996;**TR-105**: 01-19

[2] Eiamsa-ard S, Promvonge P. Enhancement of Heat Transfer in a Tube With Regularly-Spaced Helical Tape Swirl Generators'. Japan: McGraw Hill Education; 2005:**72–49-34**. New York: ASME. p. 2

[3] Bergles AE. Techniques to Augment Heat Transfer, Handbook of Heat Transfer Application. Universitá di Parma, Dipartimento di Ingegneria Industriale V. le delle Scienze. New York: McGraw-Hill; 1983;**43100**:1-8

[4] Gouda R, Das AB. Some Experimental Studies on Heat Transfer Augmentation. Rourkela: Department of Chemical Engineering National Institute of Technology; 2008. pp. 4-5

[5] Bergles AF, Lee RA, Mike BB. Heat transfer in rough tubes with tape generated swirl flow. Journal of Heat Transfer, ASME, Transaction. 1969;**8**: 443-445

[6] Shah RK, London AL. Laminar Flow Forced Convection in Ducts. New York: Academic Press; 1978. pp. 379-384

[7] Ljubicic B. Testing of Twisted-Tube Exchangers in Transition Flow Regime. International Conference on Compact Heat Exchangers and Enhancement Technology for the Process Industries. Canada; 1999

[8] Ahulkumar AR, Srinivas PN. Experimental analysis of heat transfer augmentation by using twisted tapes of different twist ratio as flow arrangement inside the tubes. IJMER. 2013;**3**(3):1317-1323

[9] Chaudhari NR, Adroja FN. "A review on design & analysis of double pipe heat exchanger". International Journal of Engineering Research & Technology (IJERT).2014;**3**(2):2502-2505

[10] Dehankar PB, Patil NS. Heat transfer augmentation – a review for helical tape insert. International Journal of Scientific Engineering and Technology. 2014;**3**(10):1236-1238

[11] Kern DQ. Process Heat Transfer. McGraw Hill Education; 1997. pp. 62-127

[12] Sara Rainieri, Angelo Farina and Giorgio Pagliarini.' Experimental investigation of heat transfer and pressure drop augmentation for laminar flow in spirally enhanced tube', paper.

Chapter 7

Heat Exchanger Design and Optimization

Shahin Kharaji

Abstract

A heat exchanger is a unit operation used to transfer heat between two or more fluids at different temperatures. There are many different types of heat exchangers that are categorized based on different criteria, such as construction, flow arrangement, heat transfer mechanism, etc. Heat exchangers are optimized based on their applications. The most common criteria for optimization of heat exchangers are the minimum initial cost, minimum operation cost, maximum effectiveness, minimum pressure drop, minimum heat transfer area, minimum weight, or material. Using the data modeling, the optimization of a heat exchanger can be transformed into a constrained optimization problem and then solved by modern optimization algorithms. In this chapter, the thermal design and optimization of shell and tube heat exchangers are presented.

Keywords: log-mean temperature difference (LMTD), effectiveness-number of transfer units (ε-NTU), genetic algorithm (GA), particle swarm optimization (PSO)

1. Introduction

Heat exchangers are systems used to transfer heat between fluids with different temperatures. These devices have vast applications in many areas, such as refrigeration, heating, and air conditioning systems, power plants, chemical processes, food industry, automobile radiators, and waste heat recovery units. Heat exchangers can be classified according to different criteria such as construction, flow arrangement, heat transfer mechanism, etc [1]. The heat exchanger design can be divided into two main categories, thermal and hydraulic design and mechanical design. In thermal and hydraulic design, the focus is on calculating an adequate surface area transfer a certain amount of heat, pressure dope, pumping power work, etc. The goal of the mechanical design is to design the mechanical integrity of the exchanger, as well as designing various pressure and non-pressure components. In this chapter, the thermal and hydraulic design of heat exchangers is presented. To achieve better performance of heat exchangers, they optimize based on their application. Heat exchanger optimization can be performed using different optimization algorithms. Since most heat exchanger optimization problems are nonlinear, using traditional methods such as linear and dynamic programming and steepest descent may not lead to the desired solution and may even fail. Also, most traditional methods need gradient information to solve an optimization problem. On the other hand, advanced optimization algorithms are developed, which are gradient-free. Several advanced optimization methods, such as genetic algorithm, non-nominated sorting genetic algorithm, bio-geography-based optimization, particle swarm optimization, Jaya algorithm, and teaching-learning-based optimization, can be more efficient in solving an optimization problem. However, each of these methods has its advantages and disadvantages, which are discussed in the optimization section. In this chapter, genetic algorithm and particle swarm optimization are discussed in detail due to the vast applications that arise from their acceptable accuracy, as well as short computational time [2].

2. Basic equation of heat transfer

In most heat transfer problems, hot and cold fluids are divided by a solid wall. In this case, the mechanism of heat transfer from hot fluid to the cold fluid can be categorized into three steps:

- Heat transfer from the hot fluid to the wall by convection.
- Heat transfer through the wall by conduction.
- Heat transfer from the wall to the cold fluid by convection.

Figure 1 shows a schematic of heat transfer between two fluids. As it can be seen, thermal resistance (R) is present at each stage of the transfer. Thermal resistance is a thermal (physical) property that indicates the resistance of each material to heat transfer due to temperature differences that can be calculated from [3]:

$$\begin{aligned} R &= \frac{L}{KA} \ , \ for \ \ conduction \\ R &= \frac{1}{hA} \ , \ for \ \ convection \end{aligned} \tag{1}$$

Where L is the thickness of the wall, A is the cross-sectional area in which heat transfer occurs, and K and h are conduction and convection heat transfer coefficient, respectively.

Heat transfer in each stage can be calculated as follows [4]:

$$Q = \frac{T_1 - T_2}{R_{c,H}} = \frac{T_2 - T_3}{R_{f,H}} = \frac{T_3 - T_4}{R_w} = \frac{T_4 - T_5}{R_{f,C}} = \frac{T_5 - T_6}{R_{c,C}} \tag{2}$$

Where:

$R_{c,H}$ = thermal resistance for convection in the hot side.

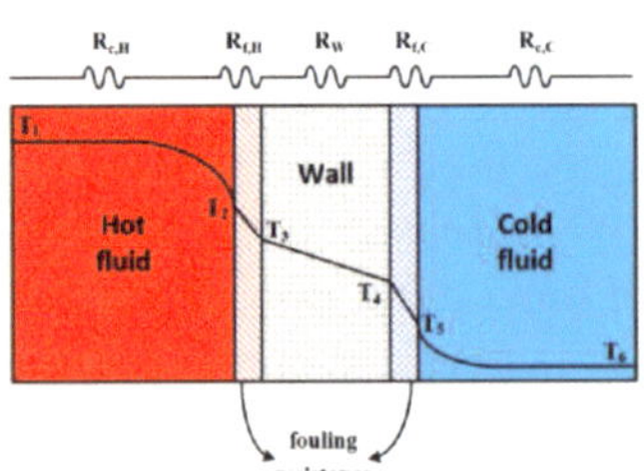

Figure 1.
A schematic of heat transfer in heat exchangers.

$R_{f,H}$ = fouling resistance of hot side.
R_w = wall resistance.
$R_{f,C}$ = fouling resistance of cold side.
$R_{c,C}$ = thermal resistance for convection in cold side.

Table 1 shows the fouling resistance of the most common fluids used in heat exchangers. The overall heat transfer coefficient can be obtained from Eq. (2) as follows [4]:

$$Q = \frac{T_1 - T_6}{R_{c,H} + R_{f,H} + R_w + R_{f,C} + R_{c,C}} \tag{3}$$

or:

$$Q = \frac{T_1 - T_6}{\frac{1}{h_H A_H} + \frac{r_{f,H}}{A_H} + \frac{r_w}{A_H} + \frac{r_{fC}}{A_C} + \frac{1}{h_C A_C}} \tag{4}$$

Gas and vapors	Fouling Factor $\left(\frac{hr-ft^2-°F}{Btu}\right)$	Liquids				Fouling Factor $\left(\frac{hr-ft^2-°F}{Btu}\right)$
Industrial		**Industrial liquids**				
Manufactured Gas	0.01	Industrial organic heat transfer media				0.001
Engine Exhaust Gas	0.01	Refrigerating liquids				0.001
Steam (non-oil bearing)	0.0005	Molten heat transfer salts				0.0005
Exhaust Steam (oil bearing)	0.001	Hydraulic fluid				0.001
Refrigerant Vapors (oil bearing)	0.002	**Industrial oils**				
Compressed Air	0.002	Fuel oil				0.005
Industrial Organic Heat Transfer Media	0.001	Engine lube oil				0.001
Chemical Processing		Transformer oil				0.001
Acid Gas	0.001	Quench oil				0.004
Solvent Vapors	0.001	Vegetable oils				0.003
Stable Overhead Product	0.001	**water**				
Petroleum Processing		Temperature of heating medium	240°F≤		240–400°F	
Atmospheric Tower Overhead Vapors	0.001	Temperature of water	125°F≤		125°F>	
Light Naphthas	0.001	velocity	3 ft.≤	3 ft.>	3 ft. ≤	3 ft. >
Vacuum Overhead Vapors	0.002	Seawater	0.0005	0.0005	0.001	0.001
Natural Gas	0.001	Brackish water	0.002	0.001	0.003	0.002
Overhead Products	0.001	Distilled water	0.0005	0.0005	0.0005	0.0005
Coke Unit Overhead Vapors	0.002	Boiler blowdown	0.002	0.002	0.002	0.002

Table 1.
Fouling factors for different types of fluid [5, 6].

Where:

h_H and h_C = convection heat transfer coefficient of the hot and cold sides, respectively.

A_H and A_C = and surface area of wall in the hot and cold side, respectively.

The r_w can be calculated for flat wall and cylindrical walls using Eqs. (5) and (6), respectively.

$$r_w = \frac{d_w}{KA}, \text{ } for \text{ } flat \text{ } wall \tag{5}$$

$$r_w = \frac{\ln\left(\frac{r_o}{r_i}\right)}{2\pi LK}, \text{ } for \text{ } cylindrical \text{ } wall \tag{6}$$

Where d_w is the thickness of the wall, and r_o and r_i are the outside and inside diameter of the wall, respectively.

Total thermal resistance can be expressed as [7]:

$$R_t = \frac{1}{h_H A_H} + \frac{r_{f,H}}{A_H} + \frac{r_w}{A_H} + \frac{r_{f,C}}{A_C} + \frac{1}{h_C A_C} \tag{7}$$

The rate of heat transfer (Q) can be determined from

$$Q = UA\Delta T \tag{8}$$

Where U is the overall heat transfer coefficient.

$$U = \frac{1}{\frac{1}{h_H \frac{A_H}{A_{ref}}} + \frac{r_{f,H}}{\frac{A_H}{A_{ref}}} + \frac{r_w}{\frac{A_H}{A_{ref}}} + \frac{r_{f,C}}{\frac{A_C}{A_{ref}}} + \frac{1}{h_C \frac{A_C}{A_{ref}}}} \tag{9}$$

Where A_{ref} is a reference area. If the heat transfer is carried out over a pipe, the inside and outside surface areas of the pipe are not equal. Hence, the A_{ref} must be determined (The outer surface of the pipes is usually selected).

3. Thermal design of heat exchangers

The thermal design of heat exchangers can be performed by several methods. The most commonly used methods are log-mean temperature difference (LMTD) and effectiveness-number of transfer units (ε-NTU) [8]. The LMTD was used to calculate heat transfer when the inlet and outlet temperatures of fluids are specified. When more than one inlet and/or outlet temperature of the heat exchanger is unknown, LMTD may be calculated by trial and errors solution. In this case, the ε-NTU method is commonly used [3].

3.1 The log-mean temperature difference (LMTD) method

As mentioned earlier, by determining the temperature difference between hot and cold fluids, the amount of heat transfer can be calculated from Eq. (3). **Figure 2** shows temperature changes of hot and cold fluids along with a heat exchanger with different types of flow configuration. As it can be seen the temperature difference between hot and cold fluids can vary along with the heat exchanger. Terminal temperatures of hot and cold fluids ($T_{H,out}$ anf $T_{C,out}$) are very effective factors in a heat exchanger design. If $T_{C,out}$ is lower than $T_{H,out}$ for countercurrent flow,

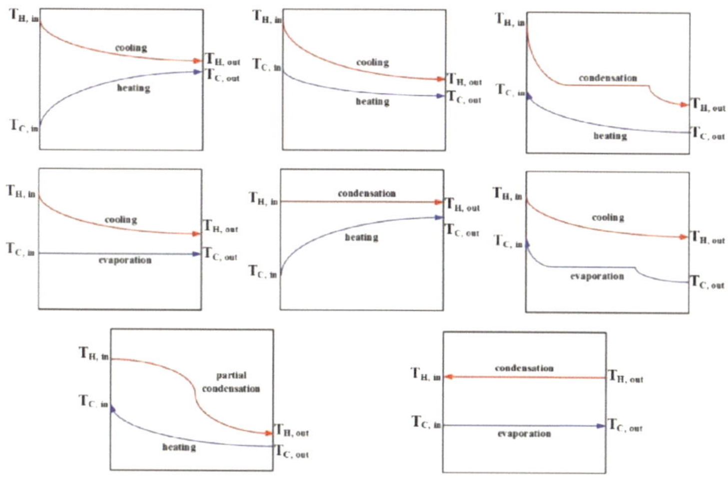

Figure 2.
Temperature changes of hot and cold fluids along with a heat exchanger with different types of flow configuration.

temperature approach occurs. In contrast, if $T_{C,out}$ is higher than the $T_{H,out}$ for countercurrent flow, temperature cross happens [9]. But if $T_{C,out}$ is equal to $T_{H,out}$, temperature meet takes place. Based on the second law of thermodynamics, temperature cross can never take place for heat exchangers with co-current flow configuration [10]. In 1981, Wales [11] proposed a parameter G, which can be used to determine the temperature conditions in the heat exchangers. Eq. (10) defines the G parameters that can be between −1 and 1.

$$
\begin{aligned}
&G = \frac{T_{H,out} - T_{C,out}}{T_{H,in} - T_{C,in}}, \\
&G > 0 \rightarrow \ temperature \ \ approach \\
&G = 0 \rightarrow temperature \ \ meet \\
&G < 0 \rightarrow temperature \ \ cross
\end{aligned} \tag{10}
$$

Since the temperature difference between hot and cold streams varies along with the heat exchanger, the basic question is which temperature difference should be considered to calculate the heat transfer rate. To answer this question, consider **Figure 3** which shows heat transfer between two parallel and co-current fluids. Based on **Figure 3**, the heat transfer for the specified heat transfer area can be written in the form:

$$dQ = U \ \Delta T \ dA \tag{11}$$

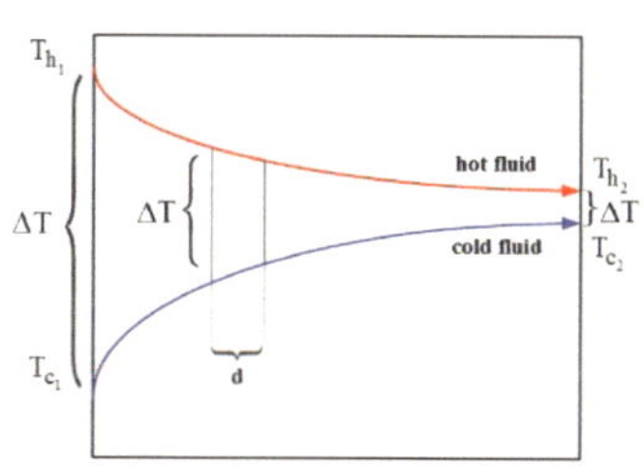

Figure 3.
Heat transfer between two parallel and co-current fluids [3].

It can be said that the amount of heat transferred is reduced from the hot fluid and added to the cold fluid. Therefore [3]:

$$dQ = -C_{p,H}\ dT_H \tag{12}$$

$$dQ = C_{p,C}\ dT_C \tag{13}$$

Where $C_{p,H}$ and $C_{p,C}$ are specific heat capacities of hot and cold fluids, respectively. The temperature difference can be written as below:

$$d(\Delta T) = dT_H - dT_C \tag{14}$$

By the combination of Eqs. (12) and (13) with Eq. (14):

$$d(\Delta T) = -\frac{dQ}{C_{p,H}} - \frac{dQ}{C_{p,C}} \tag{15}$$

By defining $\frac{1}{M} = \frac{1}{C_{p,C}} + \frac{1}{C_{p,H}}$, Eq. (15) can be written as follow:

$$dQ = M\ d(\Delta T) \tag{16}$$

Assuming the M is constant along with the heat exchanger:

$$\int_0^Q dQ = -M \int_{\Delta T_{in}}^{\Delta T_{out}} d(\Delta T) \rightarrow Q = -M(\Delta T_{out} - \Delta T_{in}) \tag{17}$$

On the other hand, by placing Eq. (16) in Eq. (11):

$$-M\frac{d(\Delta T)}{\Delta T} = UdA = A\frac{UdA}{A} \tag{18}$$

$$A\int_0^A \frac{UdA}{A} = -M \int_{\Delta T_{in}}^{\Delta T_{out}} \frac{d(\Delta T)}{\Delta T} \rightarrow U_m A = -M\ \ln\left(\frac{\Delta T_{out}}{\Delta T_{in}}\right) \tag{19}$$

Where U_m is the mean overall heat transfer coefficient which can be defined as below:

$$U_m = \int_0^A \frac{U\ dA}{A} \tag{20}$$

The heat transfer rate can be written as flow:

$$Q = U_m A \Delta T_{LMTD} \tag{21}$$

Where ΔT_{LMTD} is the logarithmic mean temperature difference (LMTD) which can be defined as below:

$$\Delta T_{LMTD} = \frac{\Delta T_{out} - \Delta T_{in}}{\ln\left(\frac{\Delta T_{out}}{\Delta T_{in}}\right)} \tag{22}$$

The simplicity of the LMTD method has led to its use in the design of many heat exchangers by introducing a correction factor, F, according to Eq. (23) [12]. The F is generally expressed in terms of two non-dimensional parameters, thermal effectiveness (P), and heat capacity ratio (R). The P and R are defined as Eqs. (24) and (25), respectively [13]. **Figure 4** shows the correction factor for common shell and tube heat exchangers.

$$Q = UAF\Delta T_{LMTD} \ , \ 0 < F < 1 \tag{23}$$

$$P = \frac{T_{C,out} - T_{C,in}}{T_{H,in} - T_{C,in}} \tag{24}$$

$$R_{cr} = \frac{T_{H,in} - T_{H,out}}{T_{C,out} - T_{C,in}} \tag{25}$$

3.2 Effectiveness-number of transfer units (ε-NTU)

When more than one of the inlet and outlet temperatures of the heat exchanger is unknown, LMTD may be calculated by trial and errors solution. Another approach to calculating the rate of heat transfer is the effectiveness number of transfer units (ε-NTU) method. The ε-NTU can be expressed according to Eq. (26) where Cp_{min} is the minimum value between the heat capacity of cold fluid ($Cp,_C$) and hot fluid ($Cp,_H$). The effectiveness (ε) can be defined as the ratio of the actual heat transfer rate (q) and the maximum possible heat transfer rate (q_{max}) according to Eq. (27).

$$NTU = \frac{UA_m}{Cp_{\min}} \tag{26}$$

$$\varepsilon = \frac{q}{q_{\max}} \tag{27}$$

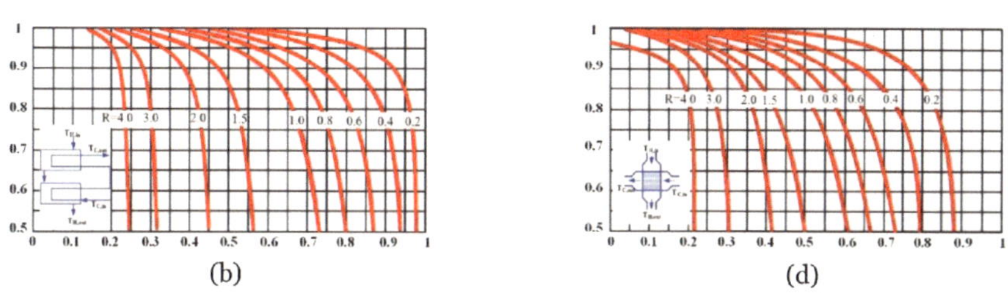

Figure 4.
Correction factor for common shell and tube heat exchangers [3]. (a) One-shell pass and 2, 4, 6, etc. (any multiple of 2), tube pass. (b) Two-shell pass and 4, 8, 12, etc. (any multiple of 4), tube pass. (c) Single-pass cross-flow with both fluids unmixed. (d) Single-pass cross-flow with one fluid unmixed and other unmixed.

Where:

$$q = Cp_{,H}(T_{H,in} - T_{H,out}) = Cp_{,C}(T_{C,out} - T_{C,in}) \tag{28}$$

$$q_{\max} = C_{p\min}(T_{H,in} - T_{C,in}) \tag{29}$$

The heat transfer rate using the ε-NTU method can express as [14]:

$$Q = \varepsilon C_{P\min}(T_{H,in} - T_{C,in}) \tag{30}$$

Tables 2 and **3** show the effectiveness and NTU relations for heat exchangers, respectively. It should be noted that C_r is the capacity and it can be defined as follows:

$$C_r = \frac{C_{P\min}}{C_{P\max}} \tag{31}$$

Heat exchanger type	Effectiveness relation
Double pipe: • Parallel-flow • Counter flow with $C_r < 1$ • Counter flow with $C_r = 1$	$\varepsilon = \frac{1-\exp[-NTU(1+C_r)]}{1+C_r}$ $\varepsilon = \frac{1-\exp[-NTU(1-C_r)]}{1-C_r\exp[-NTU(1-C_r)]}$ $\varepsilon = \frac{NTU}{1+NTU}$
Sell and tube: • On-pass and 2, 4, ... tube	$\varepsilon = 2\left\{1 + C_r + \sqrt{1+C_2}\frac{1+\exp\left[-NTU\sqrt{1+C_2}\right]}{1-\exp\left[-NTU\sqrt{1+C_2}\right]}\right\}^{-1}$
Cross flow (single-pass): • C_{max} and C_{min} unmixed • C_{max} mixed and C_{min} unmixed • C_{max} unmixed and C_{min} mixed	$\varepsilon = 1 - \exp\left[\frac{1}{C_r}(NTU)^{0.22}\left\{\exp\left[-C_r(NTU)^{0.78}\right] - 1\right\}\right]$ $\varepsilon = \frac{1}{C_r}\{1 - \exp[-C_r(1 - \exp[NTU])]\}$ $\varepsilon = 1 - \exp\left\{-\frac{1}{C_r}(1 - \exp[-C_rNTU])\right\}$
All heat exchangers with $C_r = 0$	$\varepsilon = 1 - \exp(-NTU)$

Table 2.
Effectiveness relation for heat exchangers [15].

Heat exchanger type	NTU relation
Double pipe: • Parallel-flow • Counter flow with $C_r < 1$ • Counter flow with $C_r = 1$	$NTU = -\frac{\ln[1-\varepsilon(1+C_r)]}{1+C_r}$ $NTU = \frac{1}{C_r-1}\ln\left(\frac{\varepsilon-1}{\varepsilon C_r-1}\right)$ $NTU = \frac{\varepsilon}{1-\varepsilon}$
Sell and tube: • On-pass and 2, 4, ... tube	$NTU = -\frac{1}{\sqrt{1+C^2}}\ln\left[\frac{\frac{2}{\varepsilon}-1-C_r-\sqrt{1+C^2}}{\frac{2}{\varepsilon}-1-C_r+\sqrt{1+C^2}}\right]$
Cross flow (single-pass): • C_{max} mixed and C_{min} unmixed • C_{max} unmixed and C_{min} mixed	$NTU = -\ln\left[1 + \frac{\ln(1-\varepsilon C_r)}{C_r}\right]$ $NTU = -\frac{\ln[C_r\ln(1-\varepsilon)+1]}{C_r}$
All heat exchangers with $C_r = 0$	$NTU = -\ln[1 - \varepsilon]$

Table 3.
NTU relation for heat exchangers [15].

4. Thermal and hydraulic design of shell and tube heat exchanger

Heat exchangers can be classified according to different criteria such as construction, flow arrangement, heat transfer mechanism, etc [1]. Shell and tube heat exchangers are some of the most convenient heat exchangers due to their versatility, wide operating range, and simplicity [16]. Hence, this chapter focuses on the design of this type of heat exchanger. In the design of shell and tube heat exchangers, a lot of consideration including the number of shells and tubes, tube pitch and layout, tube passes, baffles, etc., should be taken into account. In this case, there are some methods such as Kern and Bell-Delaware to design a heat shell and tube exchanger design. Since Kern's method offers the simplest route, this chapter is focused on this method.

4.1 Kern's method

Kern's method is based on experimental data for typical heat exchangers. In this method, it is assumed the shell flow is ideal, and leakage and bypass are negligible. Based on this flow model, only a single stream flows in the shell that is driven by baffles. This can lead to a very simple and rapid calculation of shell-side coefficients as well as pressure drop [17]. **Figure 5** shows a schematic of a shell and tube heat exchanger.

4.2 Number of tubes

The number of tubes (N_t) can be calculated as follows:

$$N_t = \frac{4\,\dot{m}_t}{\rho_t \nu_t \pi d_i^2} \tag{32}$$

Where $\dot{m}_t$ is the flow rate of fluid inside the tube, ρ_t is the density of the fluid inside the tube, ν_t is the velocity of the fluid inside the tube, A_t is the cross-sectional area of the tube, and d_i is the tube inside diameter.

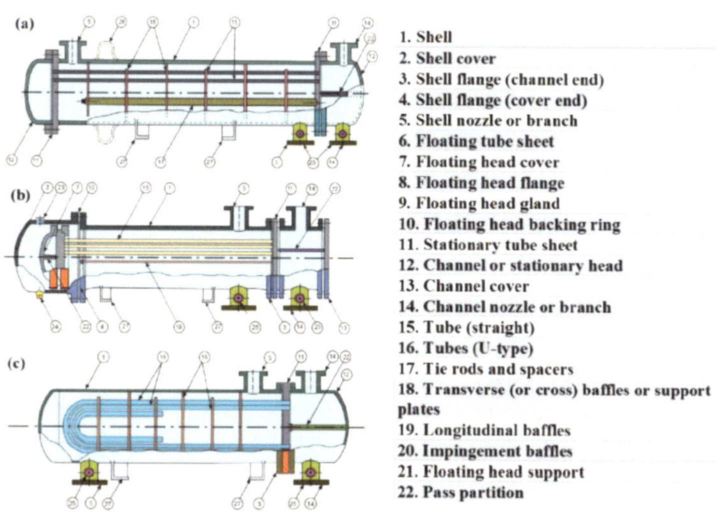

Figure 5.
Schematic of a shell and tube heat exchanger a) fixed-tube b) floating-head c) removable U-tube [18].

4.3 Tube-side heat transfer coefficient

The heat transfer coefficient for the tube side (h_t) is calculated as follows:

$$h_t = Nu_t \frac{k_t}{d_i} \tag{33}$$

Where Nu_t is the Nusselt number for the tube-side fluid and k_t is the thermal conductivity of the tube-side fluid. The Nu_t is a function of Reynolds number (Re) and Prandtl number (Pr). Re and Pr can be obtained by the following:

$$\mathrm{Re}_t = \frac{\rho_t v_t d_t}{\mu_t} \tag{34}$$

$$\mathrm{Pr}_t = \frac{C_p \mu_t}{K} \tag{35}$$

Where μ_t is the dynamic viscosity of the tube-side fluid, K is the heat conductivity coefficient, and C_p is the heat capacity of the tube-side fluid. The Nu_t can be calculated according to the type of flow as follows:

$$Nu_t = \frac{(f_t/2)\,\mathrm{Re}_t \mathrm{Pr}_t}{1.07 + 12.7(f_t/2)^{1/2}\left(\mathrm{Pr}_t^{2/3} - 1\right)} ;\textit{for} : 10^4 < \mathrm{Re} < 5 \times 10^6 \,\&\; 0.5 < \mathrm{Pr} < 200 \tag{36}$$

$$Nu_t = 1.86\left(\frac{\mathrm{Re}_t \mathrm{Pr}_t d_i}{L}\right)^{1/3} ;\; \textit{for} : \left(\frac{\mathrm{Re}_t \mathrm{Pr}_t d_i}{L}\right)^{1/3} > 2 \;\&\; 0.48 < \mathrm{Pr} < 16700 \tag{37}$$

Where L is the length of the tube and f_t is the friction factor of the tube side, which can be calculated from

$$f_t = (1.58 \ln \mathrm{Re}_t - 3.28)^{-2} \tag{38}$$

The convection heat transfer coefficient in the tube is obtained based on the value of the Re_t from [19]:

$$h_t = \frac{k_t}{d_i}\left[3.657 + \frac{0.0677\left(\mathrm{Re}_t \mathrm{Pr}_t \frac{d_i}{L}\right)^{1.3}}{1 + 0.1\mathrm{Pr}_t\left(\mathrm{Re}_t + \frac{d_i}{L}\right)^{0.3}}\right] ;\; \textit{for}\;\; \mathrm{Re}_t < 2300 \tag{39}$$

$$h_t = \frac{k_t}{d_i}\left[\frac{\frac{\lambda}{8}(\mathrm{Re}_t - 1000)\mathrm{Pr}_t}{1 + 12.7\sqrt{\frac{\lambda}{8}}\left(\mathrm{Pr}_t^{0.67} - 1\right)}\left(1 + \left(\frac{d_i}{L}\right)^{0.67}\right)\right] ;\; \textit{for}\;\; 2300 < \mathrm{Re}_t < 10000 \tag{40}$$

$$h_t = \frac{k_t}{d_i} 0.027\,\mathrm{Re}_t^{0.8} \mathrm{Pr}_t^{\frac{1}{3}}\left(\frac{\mu_t}{\mu_{w,t}}\right)^{0.14} ;\; \textit{for}\;\; \mathrm{Re}_t > 10000 \tag{41}$$

Where $\mu_{w,t}$ is the dynamic viscosity of the tube-side fluid at the wall temperature and λ is the Darcy friction coefficient which can be defined as [19]:

$$\lambda = \left(1.82 \log_{10} \log_{10} \mathrm{Re}_t - 1.64\right)^2 \tag{42}$$

The tube-side pressure drop is calculated by the following:

$$\Delta P_t = \left(4f_t \frac{LN_p}{d_i} + 4N_p\right) \frac{\rho_t \mu_t^2}{2} \tag{43}$$

Where N_p is the tube passes.

4.4 Shell diameter

Inside sell diameter (D_s) is calculated as follows:

$$D_s = \sqrt{\frac{4AN_t}{(CTP)\pi}} \tag{44}$$

Where A is the projected area of the tube layout expressed as an area corresponding to one tube and can be obtained from Eq. (45). Also, P_t is tube pitch and CL is the tube layout constant. **Figure 6** depicts two common tube layouts, square pitch and triangular pitch. The CTP is the tube count calculation constant that accounts for the incomplete coverage of the shell diameter by the tubes, due to necessary clearances between the shell and the outer tube circle and tube omissions due to tube pass lanes for multitude pass design [15]. Eq. (46) shows the CTP for different tube passes.

$$\begin{aligned} &A = P_t^2(CL) \\ &CL = 1 \rightarrow for\ \ square - pitch\ \ layout \\ &CL = 0.866 \rightarrow for\ \ triangular - pitch\ \ layout \end{aligned} \tag{45}$$

$$\begin{aligned} &CTP = 0.93 \rightarrow for\ \ one - tube\ \ pass \\ &CTP = 0.9 \rightarrow for\ \ two - tube\ \ pass \\ &CTP = 0.85 \rightarrow for\ \ three - tube\ \ pass \end{aligned} \tag{46}$$

Combining Eq. (44) with Eq. (45) as well as defining tube pitch ratio as P_r, one gets:

$$D_s = \sqrt{\frac{4(P_r d_o)^2 (CL) N_t}{(CTP)\pi}} \tag{47}$$

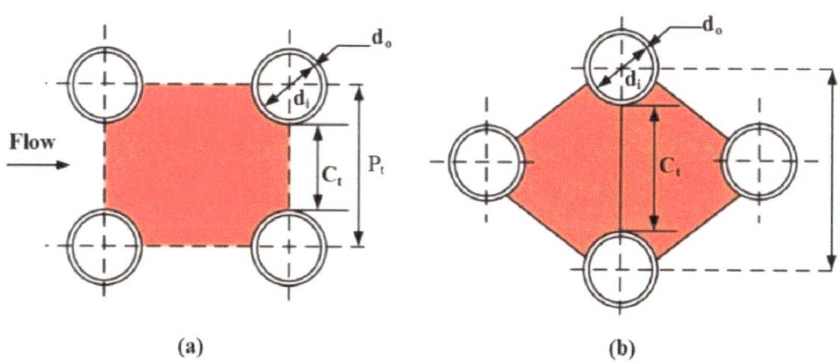

Figure 6.
(a) Square-pitch (b) triangular-pitch layout [20].

$$P_r = \frac{P_t}{d_o} \quad (48)$$

Where d_o is tube outside the diameter. Eq. (47) can be written as follows:

$$D_s = \sqrt{\frac{4(P_r^2)(CL)A_o d_o}{(CTP)\pi^2 L}} \quad (49)$$

Where A_o is the outside heat transfer surface area based on the outside diameter of the tube and can be calculated from:

$$A_o = \pi d_o N_t L \quad (50)$$

The shell side flow direction is partially along the tube length and partially across to tube length or heat exchanger axis. The inside shell diameter can be obtained based on the cross-flow direction and the equivalent diameter (D_e) is calculated along the long axes of the shell. The equivalent diameter is given as follows:

$$D_e = \frac{4 \times free - flow\ \ area}{wetted\ \ perimeter} \quad (51)$$

From **Figure 6** the equivalent diameter for the square pitch and triangular pitch layouts are as below:

$$D_e = \frac{4\left(P_t^2 - \frac{\pi d_o^2}{4}\right)}{\pi d_o}\ ;\ for\ \ square - pitch\ \ tube \quad (52)$$

$$D_e = \frac{4\left(\frac{P_t^2\sqrt{3}}{4} - \frac{\pi d_o^2}{8}\right)}{\frac{\pi d_o}{2}}\ ;\ for\ \ triangular - pitch\ \ tube \quad (53)$$

Reynolds number for the shell-side (Re_s) can be obtained as follows:

$$\mathrm{Re}_s = \left(\frac{\dot{m}_s}{A_s}\right)\frac{D_e}{\mu_s} \quad (54)$$

Where $\dot{m}_s$ is the flow rate of shell-side fluid, μ_s is the viscosity of the shell-side fluid, and A_s is the cross-flow area at the shell diameter which can be obtained as below:

$$A_s = \frac{D_s}{P_t}(B \times C_t) \quad (55)$$

Where B is the baffle spacing and C_t is the clearance between adjacent tubes. According to **Figure 6** C_t is expressed as follows:

$$C_t = P_t - d_o \quad (56)$$

The shell-side mass flow rate (G_s) is found with:

$$G_s = \frac{\dot{m}_s}{A_s} \quad (57)$$

In Kern's method, the heat transfer coefficient for the shell-side (h_s) is estimated from the following:

$$h_s = \frac{0.36k_s}{D_e}\,\mathrm{Re}_s^{0.55}\mathrm{Pr}_s^{1/3}$$
$$for\ \ 2\times 10^3 < \mathrm{Re}_s = \frac{G_s D_e}{\mu_s} < 1\times 10^6 \tag{58}$$

Where k_s is the thermal conductivity of the shell-side fluid. The tube-side pressure drop is calculated by the following:

$$\Delta P = \frac{f_s G_s^2 (N_b + 1) D_s}{2\rho_s D_e \left(\frac{\mu_b}{\mu_{w,s}}\right)^{0.14}} \tag{59}$$

Where N_b is the number of baffles, ρ_s is the density of the shell-side fluid, μ_b is the viscosity of the shell-side fluid at bulk temperature, and $\mu_{w,s}$ is the viscosity of the tube-side fluid at wall temperature. The f_s is the friction factor for the shell and can be obtained as follows:

$$f_s = \exp\left[0.576 - 0.19\ln\left(\mathrm{Re}_s\right)\right]\ ;\ for\ \ 400,\ \mathrm{Re}_s < 1\times 10^6 \tag{60}$$

The wall temperature can be calculated as follows:

$$T_w = \frac{1}{2}\left(\frac{T_{H,in} + T_{H,out}}{2} + \frac{T_{C,in} + T_{C,out}}{2}\right) \tag{61}$$

According to Eq. (21), the heat transfer surface area (A) of the shell and tube heat exchanger is obtained by the following:

$$A = \frac{Q}{U_m F \Delta T_{LMTD}} \tag{62}$$

The required length of the heat exchanger can be calculated based on the heat transfer surface area as follows:

$$L = \frac{A}{\pi d_o N_t} \tag{63}$$

5. Optimization of heat exchangers

The applications of heat exchangers are very different. Therefore, they are optimized based on their application. The most common criteria for optimizing heat exchangers are the minimum initial cost, minimum operating cost, maximum effectiveness, minimum pressure drop, minimum heat transfer area, minimum weight or material, etc. These criteria can be optimized individually or in combination. It is clear from the above that the optimal design of heat exchangers is based on many geometrical and operational parameters with high complexity. So it is difficult to design a cheap and effective heat exchanger. The optimization techniques are usually applied to ensure the best performance as well as lower the cost of the heat exchanger. The optimization is carried out using different techniques. Traditional techniques such as linear and dynamic programming as well as steepest descent usually fail to solve nonlinear large-scale problems. The need for gradient information is another drawback of traditional techniques. Therefore, it is not possible to solve non-differentiable functions using these methods. To overcome these difficulties, advanced optimization algorithms are developed which are gradient-free. Several advanced optimization methods, such as genetic algorithm (GA) [21], non-nominated sorting GA (NSGA-II) [22], bio-geography-based optimization (BBO) [23], particle swarm optimization (PSO) [24], Jaya algorithm, and teaching–learning-based optimization (TLBO) [25], had been used for the optimization of heat exchangers by many researchers each of which has its advantages and disadvantages. Using GA, it is possible to solve all optimization problems, which can be described with the chromosome encoding and solves problems with multiple solutions. But in order to use GA, it is necessary to set a number of specific algorithmic parameters such as jump probability, selection operator, cross probability. NSGA-II has an explicit diversity preservation mechanism and elitism prevents an already found Pareto optimal solution from being removed, but crowded comparison can limit the convergence and it needs the tuning of algorithmic-specific parameters including mutation probability, crossover probability, etc. The optimization using BBO is also effective and it inhibits the degradation of the solutions, but poor exploiting the solutions is the main drawback of this method. The PSO is a heuristic and derivative-free technique that has the character of memory but it needs the tuning of algorithmic specific parameters and plurality of the population is not enough to achieve the global optimal solution. Similarly, TLBO and Jaya need the tuning of their own algorithmic-specific parameters [26].

Generally, an optimization design starts by selecting criteria (quantitatively) to minimize or maximize, which is called an objective function. In an optimization design, the requirements of a particular design such as required heat transfer, allowable pressure drop, limitations on height, width and/or length of the exchanger are called constraints. Several design variables such as operating mass flow rates and operating temperatures can also participate in an optimization design [15]. The single target optimization can be expressed as [27]:

$$\begin{aligned} &\min\ f(x) \\ &g_i(x) \geq 0,\ j = 1,\ 2,\ \ldots,\ J \\ &h_k(x) \geq 0,\ k = 1,\ 2,\ \ldots,\ K \end{aligned} \tag{64}$$

Where $f(x)$ is the objective function, $g_i(x) \geq 0$ is the inequality constraint, and $h_i(x) \geq 0$ is the equality constraint. Multi-objective combination optimization can be indicated as [27]:

$$\begin{aligned} &\min\ f(x) = \left(f_1(x),\ f_2(x),\ \ldots,\ f_m(x)\right) \\ &g_i(x) \geq 0,\ j = 1,\ 2,\ \ldots,\ J \\ &h_k(x) \geq 0,\ k = 1,\ 2,\ \ldots,\ K \end{aligned} \tag{65}$$

Using the data modeling, the optimization of a heat exchanger can be transformed into a constrained optimization problem and then solved by modern optimization algorithms. In this chapter, the focus is on GA and PSO because many researchers mentioned that these algorithms lead to remarkable savings in computational time and have an advantage over other methods in obtaining multiple solutions of the same quality. So it gives more flexibility to the designer [2].

5.1 Genetic algorithm

GA is a search heuristic that is inspired by Charles Darwin's theory of survival of the fittest, which explains inferior creatures pass away and superior creatures remain [28]. In GA, sets of design variables are codified by sequences with fixed or variable lengths, similar to chromosomes or individuals in biological systems. Each chromosome is formed of several design variables, which are known as genes. In repetitive processes such as GA, each repetitive stage is a generation and a collection of solutions associated with each generation is a population. Generally, the initial population is generated randomly [29]. In GA statistical methods are used to achieve optimum points. In the process of natural selection, populations are selected based on their fitness. A new population is formed using genetic operations containing selection, crossover, mutation, etc. This cycle continues until a certain result is achieved or the stop criterion is satisfied [30]. **Figure 7** shows the flowchart of GA and the steps of binary GA are discussed below [32].

Step 1: Initialization of population

The initial population of GA includes binary numbers generated randomly which are chromosomes or GA strings, consisting of bits called genes. Actually, the initial population is the probable solution to the optimization problem. The number of gens (n_g) assigned to represent a variable in the chromosome depends on the precision ϵ and the range of the variable $[x_{min}, x_{max}]$, and is given by

$$n_g = \log_2\left(\frac{x_{\min} - x_{\max}}{\in}\right) \tag{66}$$

Step 2: Fitness evaluation

The fitness value of each GA string is examined by first determining the decoded values of the variables D, and next the corresponding real values are obtained as follows:

$$x = x_{\min} + \frac{x_{\max} - x_{\min}}{2^{n_g} - 1} \tag{67}$$

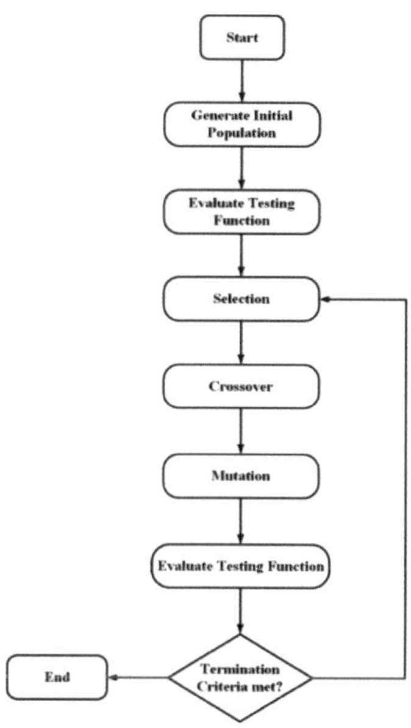

Figure 7.
Flow chart of GA [31].

The fitness function values are then computed knowing the real values of design variables.

Step 3: Reproduction/selection

In this step, chromosomes with better fitness values to participate in the crossover are selected. Several selection modes such as roulette wheel selection or proportionate selection, rank-based selection, and tournament selection can be used in this step [33]. In proportionate selection, the probability of a chromosome to be selected is directly proportional to its fitness value. Hence, the chromosome having a better fitness value has a higher chance of selection for reproduction. This may result in premature convergence of the solution because there is a chance of losing diversity. The tournament selection is faster compared with the other two selection methods. In this method, *n* chromosomes are randomly picked from the population of solutions, where n represents the tournament size. The chromosome having the best fitness value is copied to the mating pool and all the n GA strings are returned to the population. This process is repeated for obtaining all the individuals of the mating pool.

Step 4: Crossover

The genes are exchanged between two-parent chromosomes in the crossover step, which leads to a new set of solutions, called children. The crossover operation represents the selection pressure or exploitation of fit chromosomes for even better solutions. The crossover probability (P_c) specifies the number of individuals taking part in the crossover operation, and this control parameter value is optimally chosen as nearly equal to 1.0. Several schemes of crossover such as single-point crossover, two-point crossover, multipoint crossover, and uniform crossover can be used in this step. A comparison of these methods is given in the literature [34].

Step 5: Mutation

Mutation means the change of a bit from 0 to 1 and from 1 to 0 in the solution chromosome, which is used for the exploration of new solutions. It helps to come out of the local basin and search for a global solution. The mutation probability P_m specifies the number of mutations and is commonly kept very low. Because if its value is high, the qualified solutions may be lost. The range of P_m is given as

$$\frac{0.1}{l} \leq P_m \leq \frac{1}{l} \tag{68}$$

Where l represents the length of the GA string. Steps 2, 3, 4, and 5 are repeated until the termination criterion (maximum number of generations or desired precision of solution) is met.

5.2 Particle swarm optimization

The PSO is inspired by the way fish and birds swarm search for food [35]. In this method, each particle represents one solution to a problem and they aim to find optimum points in a search space. This method is also based on the behavior of birds that they use to find their orientation. Based on this direction, the collective location of the swarm and the best individual location of particles per time are calculated and a new search orientation is composed of these two orientations and the previous orientation. In a search space of the D dimension, the best individual location of a particle and the best location of the overall particle are defined as Eqs. (69) and (70), respectively.

$$\vec{P}_1 = C_1\left(\vec{P}_{i1},\ \vec{P}_{i2},\ \dots,\ \vec{P}_{iD}\right) \tag{69}$$

$$\vec{g}_1 = C_1\left(\vec{g}_1,\ \vec{g}_2,\ \dots,\ \vec{g}_D\right) \tag{70}$$

The best location in the vicinity of each particle is given as below:

$$\vec{n} = \left(\vec{n}_{i1},\ \vec{n}_{i2},\ \dots,\ \vec{n}_{iD}\right) \tag{71}$$

Displacement of particles after determination their velocity is as follows:

$$\vec{x}(t) = \vec{x}(t-1) + \vec{v}(t) \tag{72}$$

$$\vec{v}(t) = \vec{v}(t-1) + \vec{F}(t-1) \tag{73}$$

The best individual location of the particle and the best collective location of particles as two springs connected to the particle are used to model the force entered in the particle. The first spring is directed to the best individual experience and the second spring is directed to the best swarm experience. Eq. (74) shows the force entered in the particle.

$$\vec{F}_{i-1} = C_1\left(\vec{P}_{i-1} - \vec{x}_{i-1}\right) + C_2\left(\vec{g}_{i-1} - \vec{x}_{i-1}\right) \tag{74}$$

Where C_1 and C_2 are acceleration coefficients. The particle velocity at dimension $d\ (v_{id})$ and the next repetition can be obtained as follows [35]:

$$\begin{aligned} v_{id}(t) = {} & \omega v_{id}(t-1) + C_1 rand_1(P_{id}(t-1) - x_{id}(t-1)) \\ & + C_2 rand_2(n_{id}(t-1) - x_{id}(t-1)) \end{aligned} \tag{75}$$

This shows the velocity of particle i at the star topology or global best. The $rand_1$ and $rand_2$ are random numbers that have a constant distribution in the range 0–1. **Figure 8** shows the flowchart of PSO and the steps of PSO are discussed below [32].

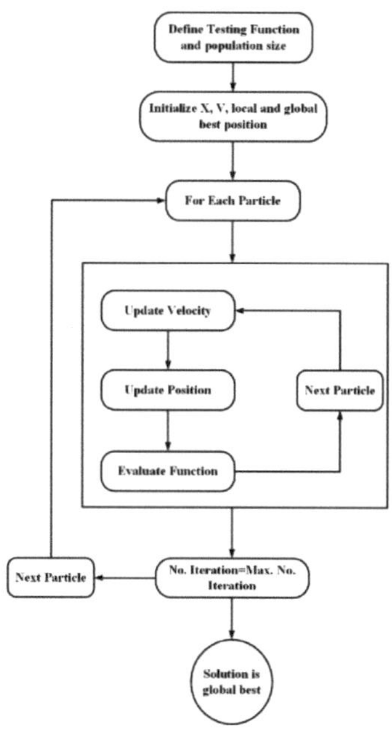

Figure 8.
Flow chart of PSO [31].

Step 1: Initialization

The swarm of potential solutions is generated with random positions and velocities. The *ith* particle in D-dimensional space may be denoted as $X_i = (x_{i1}, x_{i2},\ldots, x_{id})$ and $i = 1, 2,\ldots, N$, where N denotes the size of the swarm.

Step 2: Fitness evaluation

The corresponding fitness values of the particles are evaluated.

Step 3: Determination of personal and global best

The best individual location of a particle ($\vec{P}_1$) is sorted, the particle having the best fitness value is determined for the current generation, and the best location ($\vec{g}_1$) is updated.

Step 4: Velocity and position update

The velocity and position of the ith particle are updated based on Eq. (75).

Here, an example of the design and optimization of a shell and tube heat exchanger is presented. This example was used by Karimi et al. (2021) [36]. Their aim was to minimize the total annual cost (C_{tot}) for a shell and tube heat exchanger based on optimization algorithms. The total annual cost is the sum of the initial cost for the construction (C_i) of the heat exchanger and the cost of power consumption in the shell and tube heat exchanger (C_{od}). Hence, the total annual cost was considered as an objective function that should be minimized using GA and PSO. Process input data and physical properties for this case study are presented in **Table 4**. Also, bounds for design parameters are listed in **Table 5**. The objective function can be written as follows:

$$C_{tot} = C_i + C_{od} \tag{76}$$

The results show that the use of PSO has been led to lower Ctot, which means that the minimization of cost function was performed better using this algorithm. Also, the use of PSO resulted in lower Δp and A as well as higher U (**Table 6**).

	$\dot{m}$ $\left(\frac{kg}{s}\right)$	T_{in} (°C)	T_{out} (°C)	ρ $\left(\frac{kg}{m^3}\right)$	C_p $\left(\frac{J}{kg}\right)$	μ (Pas)	k $\left(\frac{W}{m.K}\right)$	R_f $\left(\frac{m^2K}{W}\right)$	μ_w (Pas)
Shell side: methanol	27.8	95	40	750	2840	0.00034	0.19	0.00033	0.00038
Tube side: sea water	68.9	25	40	995	4200	0.0008	0.59	0.0002	0.00052

Table 4.
Process input data and physical properties for three case studies [36].

Parameters	Lower value	Upper value
Tubes outside diameters(m)	0.015	0.051
Shell diameters(m)	0.1	1.5
Central baffle spacing(m)	0.05	0.5

Table 5.
Bounds for design parameters [36].

	PSO	GA
$L\ (m)$	2.6871	3.9089
$d_o\ (m)$	0.015063	0.015
$B\ (m)$	0.49967	0.49989
$D_s\ (m)$	0.81143	0.74105
$N_t\ (m)$	1238	1365.5
$v_t\ (m/s)$	0.83349	0.893
Re_t	27,909	13,386
Pr_t	5.69	5.69
$h_t\ (W/m^2K)$	3740.8	4639.7
f_t	0.0212	0.0073
$\Delta P_t\ (Pa)$	4730	5191.3
$D_e\ (m)$	0.0141	0.0106
$v_s\ (m/s)$	0.498	0.499
Re_s	15,489	11,716
Pr_s	5.1	5.1
$h_s\ (W/m^2K)$	9075.7	1648.9
f_s	0.313	0.353
$\Delta P_s\ (Pa)$	21,355	18,033
$U\ (W/m^2K)$	900.98	686.71
$A\ (m^2)$	198.78	252.58
$C_i\ (\$)$	44,116	50,737
$C_o\ (\$)$	2561.5	1085.2
$C_{od}\ (\$)$	2340	6685
$C_{tot}\ (\$)$	46,456	57422.51

Table 6.
Optimal parameter of heat exchanger using GA and PSO algorithms [36].

6. Conclusion

This chapter has discussed the thermal design and optimization of shell and tube heat exchangers. The basic equations of heat transfer were investigated and log-mean temperature difference (LMTD) and effectiveness-number of transfer units (ε-NTU) were presented. The thermal design was focused on Kern's method. In this method, it is assumed the shell flow is ideal, and leakage and bypass are negligible. Based on this flow model, only a single stream flows in the shell that is driven by baffles. This can lead to a very simple and rapid calculation of shell-side coefficients and pressure drop. The optimization of heat exchangers is presented based on the genetic algorithm (GA) and particle swarm optimization (PSO) due to the recommendation of these methods by man researchers because of quick convergence and obtaining multiple solutions.

Conflict of interest

The authors declare no conflict of interest.

Nomenclature

A	*Total heat transfer area*
A_s	*Cross-flow area at the shell diameter*
B	*Baffle spacing*
Cp	*Heat capacity*
C_i	*Capital investment cost*
C_o	*Total operating cost*
C_{od}	*Total discounted operating cost*
C_r	*Capacity ratio*
C_t	*Clearance between adjacent tubes*
C_{tot}	*Total annual cost*
CL	*Tube layout constant*
CTP	*Tube count calculation constant*
d_w	*Wall thickness*
D_s	*Shell diameter*
F_s	*Friction factor for shell*
G_s	*Mass flow rate of shell-side fluid*
h	*Convection heat transfer coefficient*
k	*Thermal conductivity*
K	*Heat conductivity coefficient*
L	*Tube length*
$LMTD$	*Logarithmic mean temperature difference*
$\dot{m}_s$	*Flow rate of shell-side fluid*
N_t	*Number of tubes*
NTU	Number of heat transfer unit
P	*Thermal effectiveness*
P_r	*Tube pitch ratio*
P_t	*Tube pitch constant*
Pr	*Prandtl number*
q	*Actual heat transfer rate*
q_{max}	*Maximum possible heat transfer rate*
Q	*Heat transfer rate*
r_i	*Inside diameter*
r_o	*Outside diameter*
R	*Thermal resistance*
Re	*Reynolds number*
T	*Temperature*
U	*Overall heat transfer coefficient*
U_m	*Mean overall heat transfer coefficient*

Greek symbols

Δ	Difference
ε	*Effectiveness*
ρ	*Density*
μ	*Dynamic viscosity*
ν	***Velocity***

Subscripts

b	*Bulk*
C	*Cold*

c	*Convection*
H	*Cold*
i	*Inside*
max	*Maximum*
min	*Minimum*
o	*Outside*
s	*Shell side*
t	*Tube side*

Author details

Shahin Kharaji
Research and Develop Department, FAPKCO Engineering Group, Shiraz, Iran

*Address all correspondence to: shahinkharaji@gmail.com

References

[1] Ghalandari, M., et al., Applications of intelligent methods in various types of heat exchangers: a review. Journal of Thermal Analysis and Calorimetry, 2021. 145(4): p. 1837-1848

[2] Lahiri SK, Khalfe NM,Wadhwa SK. Particle swarm optimization technique for the optimal design of shell and tube heat exchangers. Chemical Product and Process Modeling. 2012;7(1), Article 14.

[3] Ezgi CN. Basic Design Methods of Heat Exchanger. In: Heat Exchangers-Design, Experiment and Simulation. Rijeka: IntechOpen; 2017

[4] Chen Q, Ren J. Generalized thermal resistance for convective heat transfer and its relation to entransy dissipation. Chinese Science Bulletin. 2008;**53**(23): 3753-3761

[5] Marner W, Suitor J. Survey of Gas-Side Fouling in Industrial Heat-Transfer Equipment. Final Report. Pasadena, CA (USA): Jet Propulsion Lab; 1983

[6] Whitaker S. Fundamental Principles of Heat Transfer. Oxford, United kingdom: Pergamon press, Elsevier; 2013

[7] Çengel YA, Ghajar AJ. Heat and Mass Transfer: Fundamentals [and] Applications. New York City, United States: McGraw-Hill Education; 2020

[8] Borjigin S et al. Coupling ε-NTU method for thermal design of heat exchanger in cabinet cooling system. Applied Thermal Engineering. 2019;**159**: 113972

[9] Smith R. Chemical Process Design And Integration. New York City, United States: John Wiley & Sons; 2005

[10] Vengateson U. Design of multiple shell and tube heat exchangers in series: E shell and F shell. Chemical Engineering Research and Design. 2010;**88**(5–6): 725-736

[11] Wales, R. and W. RE, Mean Temperature Difference in Heat Exchangers. New York, United state Chemical Engineering Journal; 1981

[12] Fakheri A. Log mean temperature correction factor: an alternative representation. In: Proceedings of the International Mechanical Engineering Congress and Exposition. New Orleans, Louisiana: Citeseer; 2002

[13] Gulyani BB, Jain A, Kumar S. Optimal synthesis of multipass heat exchanger without resorting to correction factor. International Journal of Mechanical, Aerospace, Industrial, Mechatronic and Manufacturing Engineering. 2011;**5**(5):898-904

[14] Iu I et al. Applying the effectiveness-NTU method to elemental heat exchanger models. ASHRAE Transactions. 2007;**113**(1):504-513

[15] Shah RK, Sekulic DP. Fundamentals of Heat Exchanger Design. New York City, United States: John Wiley & Sons; 2003

[16] Silaipillayarputhur K, Khurshid H. The design of shell and tube heat exchangers–A review. International Journal of Mechanical and Production Engineering Research and Development. 2019;**9**(1):87-102

[17] Alperen MA, Kayabaşi E, Kurt H. Detailed comparison of the methods used in the heat transfer coefficient and pressure loss calculation of shell side of shell and tube heat exchangers with the experimental results. Energy Sources, Part A: Recovery, Utilization, and Environmental Effects. 2019:1-20

[18] Standard I. Specification for Shell and Tube Type Heat Exchangers. New Delhi: BIS 2007; 2007

[19] Jalilirad S, Cheraghali MH, Ashtiani HAD. Optimal design of shell-

and-tube heat exchanger based on particle swarm optimization technique. Journal of Computational Applied Mechanics. 2015;**46**(1):21-29

[20] Anescu, G., A DMSACO Approach to Economic Heat Exchanger Design. Politechnica University of Bucharest in Romania.

[21] Guo J, Cheng L, Xu M. Optimization design of shell-and-tube heat exchanger by entropy generation minimization and genetic algorithm. Applied Thermal Engineering. 2009;**29**(14–15):2954-2960

[22] Yin Q et al. Optimization design and economic analyses of heat recovery exchangers on rotary kilns. Applied Energy. 2016;**180**:743-756

[23] Hadidi A, Nazari A. Design and economic optimization of shell-and-tube heat exchangers using biogeography-based (BBO) algorithm. Applied Thermal Engineering. 2013;**51**(1–2):1263-1272

[24] Patel V, Rao R. Design optimization of shell-and-tube heat exchanger using particle swarm optimization technique. Applied Thermal Engineering. 2010;**30** (11–12):1417-1425

[25] Rao RV, Patel V. Multi-objective optimization of heat exchangers using a modified teaching-learning-based optimization algorithm. Applied Mathematical Modelling. 2013;**37**(3): 1147-1162

[26] Rao RV et al. Design optimization of heat exchangers with advanced optimization techniques: A review. Archives of Computational Methods in Engineering. 2020;**27**(2):517-548

[27] Yao J. A review of industrial heat exchange optimization. In: IOP Conference Series: Earth and Environmental Science. JIOP Publishing Ltd in Bristol, England; 2018

[28] Haupt RL, Haupt SE. Practical Genetic Algorithms. New York City, United States: John Wiley & Sons; 2004

[29] Najafi H, Najafi B, Hoseinpoori P. Energy and cost optimization of a plate and fin heat exchanger using genetic algorithm. Applied Thermal Engineering. 2011;**31**(10):1839-1847

[30] Konak A, Coit DW, Smith AE. Multi-objective optimization using genetic algorithms: A tutorial. Reliability Engineering & System Safety. 2006; **91**(9):992-1007

[31] Sadeghzadeh H, Ehyaei M, Rosen M. Techno-economic optimization of a shell and tube heat exchanger by genetic and particle swarm algorithms. Energy Conversion and Management. 2015;**93**: 84-91

[32] Swayamsiddha S. Bio-inspired algorithms: principles, implementation, and applications to wireless communication. In: Nature-Inspired Computation and Swarm Intelligence. Cambridge, Massachusetts, United States: Academic Press; 2020. pp. 49-63

[33] Blickle T, Thiele L. A Comparison of Selection Schemes used in Genetic Algorithms , TIK-Report No. 11. Computer Engineering and Communication Networks Lab (TIK) Zurich, Switzerland; 1995

[34] Spears WM, Anand V. A study of crossover operators in genetic programming. In: International Symposium on Methodologies for Intelligent Systems. Berlin, Heidelberg: Springer; 1991

[35] Shi Y. Particle swarm optimization: developments, applications and resources. In: Proceedings of the 2001 Congress on Evolutionary Computation (IEEE Cat. No. 01TH8546). Seoul, Korea (South); 2001

[36] Karimi, H., H. Ahmadi Danesh Ashtiani, and C. Aghanajafi, Optimization Techniques for Design a shell and tube heat exchanger from economic view. AUT Journal of Mechanical Engineering, 2021. **5**(2): 9-9

Chapter 8

A Review on Convective Boiling Heat Transfer of Refrigerants in Horizontal Microfin-Tubes: A Typical Example

Thanh Nhan Phan and Van Hung Tran

Abstract

Understanding the Heat transfer performance of refrigerant for convective boiling in horizontal microfin tube and smooth tube is place an importance role on the designing of evaporator, the main equipment on refrigeration system. Reviewing the general concept especially the theory of boiling in the tube, the formation of the flow pattern map, the calculating procedure for heat transfer coefficient and pressure drop during boiling process of refrigerant in microfin tube. Besides, a typical example will be presented more detail in step by step to define the heat transfer coefficient and pressure drop for one working condition to estimate the data results without doing experiments.

Keywords: convective boiling, microfin tube, flow pattern, heat transfer coefficient, pressure drop

1. Introduction

The boiling process of Refrigerant is very important on the designing of evaporator on the refrigeration system. The optimization of heat exchanger about the size, weight and heat transfer performance is the main problem. In order to solve this problem, microfin tube is the ones should pay attention.

To understand the boiling process, the pattern of flow should be known during the boiling of refrigerant. With the type of flow patterns and the general map of flow pattern for boiling in smooth tube is very clear it is concluded fully stratified, slug flow, stratified wavy flow, intermittent flow, annular flow, dryout and mist flow. While for microfin is still unclear especially for the newest refrigerant. With the method of Rollmann et al. [1], the pattern of boiling in microfin tube could be classify in such fully stratified flow, stratified wavy flow, the combination of slug and stratified wavy flow, helix flow, annular flow. In order to build this method, Rollmann et al. was updated the method of Wojtan et al. [2] which was introduced for smooth tube. Those method based on the mathematical model of two phase stratified the geometries of flow from [3] to determine the geometries of flow. Besides that, in recent years some new method to build the map for flow pattern are still updated from Zhuang et al. [4] and Yang et al. [5].

In case of internal flow, the main difficulty, concerning to the evaluation of the heat transfer coefficient and friction factor, is related to the flow regimes (distribution of liquid and vapor in a cross-section). As it is easily understood, the main heat transfer mechanism changes if, at the wall, there is liquid, vapor, both of them, droplet impingement, bubble formation and so on. Every flow regime requires a specific analytical description and appropriate criteria to state if it occurs or not. As quality, liquid mass flux and vapor mass flux change along a duct, several flow regime onsets along a duct in the presence of heat transfer. The criteria are usually represented in two-dimensional diagrams called flow pattern maps. The internal geometry of the enhanced tubes and the thermal properties of the fluid affect the distribution of the phases in the cross section.

It is obviously that the outstanding performance of heat transfer in microfin tubes for boiling is valuable to consider without introducing excessive penalization in the pressure drop. Those have been proven in so many experimental researches which are published in recent years and also presented in the paper of Phan et al. [6]. Besides, with the correlations for heat transfer of evaporation process on microfin tube are published would be introduced in this chapter are come from Thome et al. [7], Cavallini et al. [8], Yu et al. [9], Yun et al. [10], Chamra et al. [11], Wu et al. [12], Rollmann and Spindle [13].

Also for the pressure drop of two phase flow, some correlations were built for both evaporation and condensation Choi et al. [14], Goto et al. [15], the others used for separated purposed. Some particular pressure drop correlations could be mentioned for evaporation Rollmann and Spindler [13], Wongsa-Ngam et al. [16] and Kuo et al. [17].

About this problem, Phan Thanh Nhan [18] focused on the experimental test to determine flow pattern, heat transfer and pressure drop for both boiling and condensation in microfin tube for two kind of Refrigerants R134a and R1234ze.

This chapter is demonstrated in convective boiling of refrigerant in horizontal microfin tube. Flow pattern map, heat transfer coefficient and pressure drop during boiling process would be presented with the published correlations and taken on working condition for boiling as an example to calculate step by step.

2. Flow pattern map

In order to draw a map of flow mechanism, there needs to be a classification of different regimes of flow in microfin tube, which relies on flow patterns such as slug, fully stratified, stratified wavy, helix, annular and some regimes are the combination of two or three flow pattern simultaneously happened. To build the map, many parameters have been cleared about geometries of tube, heat flux, mass flux, thermo-physical and thermodynamics properties of fluid, local quality, quality change of boiling processes and define the void fraction, Martineli parameter or also estimate the shape of flow (about the liquid part and vapor part consisted on the position of tube).

Even many different group researches, they also classified the map into some main regimes: bubbly flow, plug flow, slug flow, intermittent flow, stratified flow, stratified wavy flow, annular flow, dry-out regime, mist flow or transition regime, other differences just only the difference name they called for the same regime. Flow patterns in horizontal flows are illustrated in **Figure 1**. Depend on every single group they have every different name for their classification regimes. Every regime can be described as below:

- Bubbly flow: due to buoyancy force, gas bubbles focus on the upper part of tube, and normally take place at the high mass flow rate.

- Plug flow: the individual small bubbles have coalesced to create long bubble, can call the name elongated bubble flow.

- Slug flow: at high velocity, the wave amplitude is so large which increasing and touching to the top of the tube

- Intermittent flow: could be defined instead of plug and slug flow

- Fully stratified flow: at low velocity, liquid and vapor are completely separated, interface between them is smooth

- Stratified wavy flow: the formation of wave in the interface between liquid and vapor of stratified flow when the velocity of gas rises up

- Annular flow: when the liquid forms around perimeter of the tube, vapor flows in the core separate and make the annular shape of liquid flow.

- Dry-out regime: at higher quality of vapor, thinner of annular liquid flow will be disappeared, at the top of tube becomes dry first, then gradually spread around tube to bottom

- Mist flow: liquid will be entrained to the core of gas phase as small droplets

- Transition regime: on the changing pattern regimes, it is not clear to define which pattern they are, so that the name transition regime is used.

For micro-fin tube, those the last few years, some other detailed regimes are called for flow patterns:

- Helix flow: the formation of helix flow due to the helical structure of microfin where the liquid flow helically through out

- Slug + helix flow: on the regime, slug and helix flow are spontaneously occurred

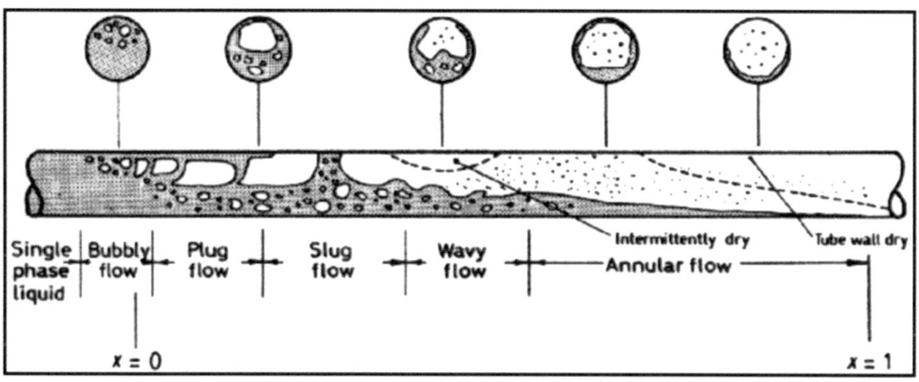

Figure 1.
Flow regime for boiling from Collier and Thome 1994 [7].

2.1 The flow pattern map of Rollmann and Spindler

Rollmann and Spindler [1] modified the procedure from Wojtan et al. [2] and introduced their method to build the flow pattern map for boiling in microfin tube as given in below (**Figure 2**):

Void fraction ε: Rouhani-Axelsson correlation

$$\varepsilon = \frac{x}{\rho_V}\left([1+0.12(1-x)]\left[\left(\frac{x}{\rho_V}\right)+\left(\frac{1-x}{\rho_L}\right)\right]+\frac{1.18(1-x)[g\sigma(\rho_L-\rho_V)]^{0.25}}{G{\rho_L}^{0.5}}\right)^{-1} \tag{1}$$

Stratified angle θstrat: Biberg correlation [19]

$$\theta_{strat} = 2\pi - 2\left\{\pi(1-\varepsilon)+\left(\frac{3\pi}{2}\right)^{1/3}\left[1-2(1-\varepsilon)+(1-\varepsilon)^{\frac{1}{3}}-\varepsilon^{\frac{1}{3}}\right]\right. \\ \left. -\frac{1}{200}(1-\varepsilon)\varepsilon[1-2(1-\varepsilon)]\left[1+4\left((1-\varepsilon)^2+\varepsilon^2\right)\right]\right\} \tag{2}$$

Geometrical parameters for two phase flow in a circular tube are showed in **Figure 3**.

$$A_{LD} = \frac{A_L}{D^2} = \frac{A(1-\varepsilon)}{D^2} \tag{3}$$

$$A_{VD} = \frac{A_V}{D^2} = \frac{A\varepsilon}{D^2} \tag{4}$$

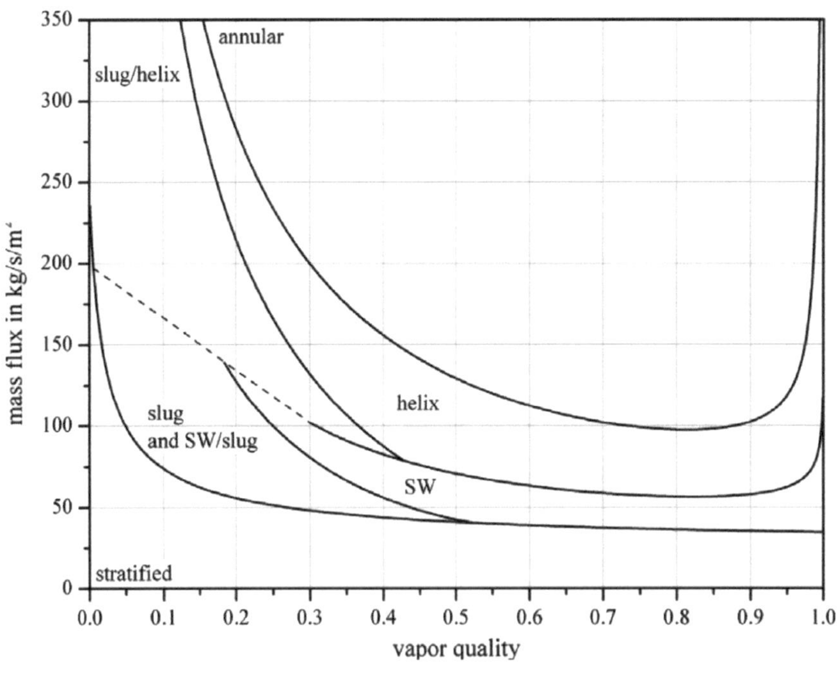

Figure 2.
Flow pattern map of Rollmann and Spindle [1].

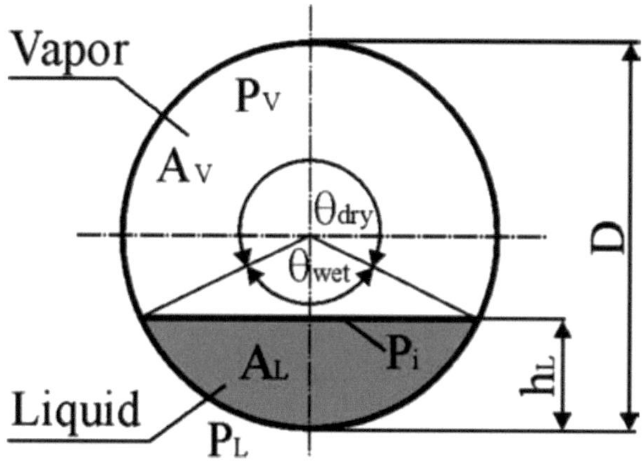

Figure 3.
Geometrical parameters of stratified flow in circular tube.

$$h_{LD} = 0.5\left(1 - \cos\left(\frac{2\pi - \theta_{strat}}{2}\right)\right) \tag{5}$$

$$P_{iD} = \sin\left(\frac{2\pi - \theta_{strat}}{2}\right) \tag{6}$$

Transition between stratified flow and stratified wavy flow:

$$G_{strat} = \left\{\frac{4\mu_L g(\rho_L - \rho_V)\rho_V \varepsilon(1-\epsilon)}{S_2 x^2(1-x)}\right\}^{1/3} + C_5; S_2 = 0.02844; C_5 = 22.9\ \text{kg/sm}^2. \tag{7}$$

Transition between slug flow and stratified wavy flow:

$$G_{slug} = \left\{\frac{16\widetilde{A}_V^3 g D\rho_L\rho_V}{x^2\pi^2\sqrt{1-(2\widetilde{H_L}-1)^2}}C_6{}^2\left[\frac{4\pi^2}{S_1{}^2\widetilde{H_L}{}^2}\left(\frac{Fr}{We}\right)_L + 1\right]\right\}^{0.5} + C_7; \left(\frac{Fr}{We}\right)_L = \frac{\sigma}{gD^2\rho_L}; S_1 = 5.889; C_6 = 1.015; C_7 = -53.35\text{kg/sm}^2, \tag{8}$$

where HL is liquid height, $\widetilde{H_L} = H_L/D$ and AV is gas phase area, $\widetilde{A_V} = A_V/D$.
Transition between stratified wavy flow and helix flow:

$$G_{SW} = \left\{\frac{16\widetilde{A}_V^3 g D\rho_L\rho_V}{x^2\pi^2\sqrt{1-\left(2\widetilde{H_L}-1\right)^2}}C_6{}^2\right\}^{0.5} + C_7, \text{for } x \geq 0.3; C_6 = 0.8441; C_7 = 0\ \text{kg/sm}^2. \tag{9}$$

For $x \geq 0.3$ the transition curve in Eq. (3) was applied. With $x < 0.3$ the linear equation With the lope of transition curve from $x \geq 0.3$.
Transition between slug/helix flow and helix flow:

$$G_{Slug-helix} = \left\{\frac{16\widetilde{A}_V^3 g D\rho_L\rho_V}{x^2\pi^2\sqrt{1-\left(2\widetilde{H_L}-1\right)^2}}C_6{}^2\right\}^{0.5} + C_7; C_6 = 1.754; C_7 = -84.79\ \text{kg/sm}^2. \tag{10}$$

Transition between helix flow and annular flow:

$$G_{slug} = \left\{ \frac{16\widetilde{A}_V^3 g D \rho_L \rho_V}{x^2 \pi^2 \sqrt{1 - \left(2\widetilde{H}_L - 1\right)^2}} C_6{}^2 \left[\frac{4\pi^2}{S_1{}^2 \widetilde{H}_L{}^2} \left(\frac{Fr}{We} \right)_L + 1 \right] \right\}^{0.5} + C_7; S_1 = 57.71; C_6 = 1.772; C_7 = -25.39\ kg/sm^2. \tag{11}$$

The classification of the regimes on the map:

Conditions	Regime
$G < G_{strat}$	Stratified flow
$G > G_{slug}$ and $G < G_{sw}$	Stratified wavy flow
$G < G_{slug}$ and $G < G_{sw}$	Stratified wavy/slug flow or slug flow
$G < G_{slug\text{-}helix}$	Slug/helix flow
$G < G_{helix}$	Helix flow
$G > G_{helix}$	Annular flow

2.2 The flow pattern map of Zhuang et al.

The flow pattern map of Zhuang et al. [4] in **Figure 4** was form from the model of Kim et al. combine with their experimental data for R170 with working range of saturation pressures from 1.5 MPa to 2.5 MPa on mass flux from 100 kg/m^{2}s to 250 kg/m^{2}s. The map was built in the terms of dimensionless weber number We and Mattinelli parameter X_{tt}.

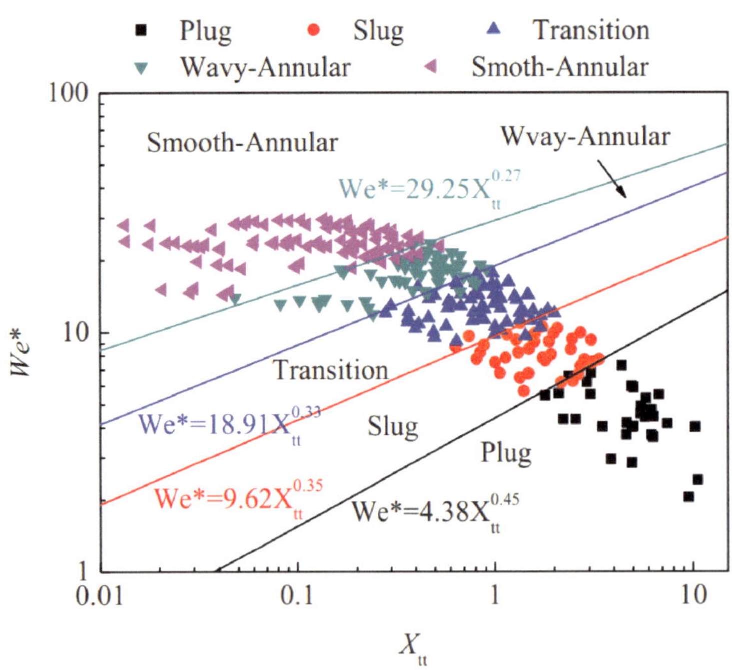

Figure 4.
Flow pattern map of Zhuang et al. [4].

Four transition curves were presented to draw the transition line to separate five different zones, each zone is named as smooth annular, wavy annular, transition, slug, plug.

Marttinelli number: X_{tt}

$$X_{tt} = \left(\frac{1-x}{x}\right)^{0.9}\left(\frac{\rho_G}{\rho_L}\right)^{0.5}\left(\frac{\mu_l}{\mu_G}\right)^{0.1} \tag{12}$$

Determine Weber number (We) based on the Reynold range of liquid flow:

$$Re_L = \frac{G(1-x)D}{\mu_L} \tag{13}$$

With $Re_L \leq 1250$

$$We^* = 2.45\frac{Re_G^{0.64}}{Su_G^{0.3}\left(1+1.09X_{tt}^{0.039}\right)^{0.4}} \tag{14}$$

With $Re_L > 1250$

$$We^* = 0.85\frac{Re_G^{0.79}X_{tt}^{0.157}}{Su_G^{0.3}\left(1+1.09X_{tt}^{0.039}\right)^{0.4}}\left[\left(\frac{\mu_G}{\mu_L}\right)^2\left(\frac{\rho_L}{\rho_G}\right)\right]^{0.084} \tag{15}$$

$$Su_G = \frac{\rho_G \sigma D}{\mu_G^2} \tag{16}$$

The transition line between patterns:

Smooth-annular to wavy-annular flow:

$$We^* = 29.25X_{tt}^{0.27} \tag{17}$$

Wavy-annular to transition flow:

$$We^* = 18.91X_{tt}^{0.33} \tag{18}$$

Transition to slug flow:

$$We^* = 9.62X_{tt}^{0.35} \tag{19}$$

Slug to plug flow:

$$We^* = 4.38X_{tt}^{0.45} \tag{20}$$

2.3 The flow pattern map of Yang et al.

The map of Yang et al. [5] introduced two maps in **Figure 5**, one map is presented for plug flow, slug flow, other is separate between slug flow and annular flow. Two transitions line on two difference maps as a function of Martinelli parameter X_{tt}, the transition lines based on three dimensionless numbers K_1, K_2, K_3 which depended on the inertia force, surface tension force, shear force, gravity force and evaporation momentum force.

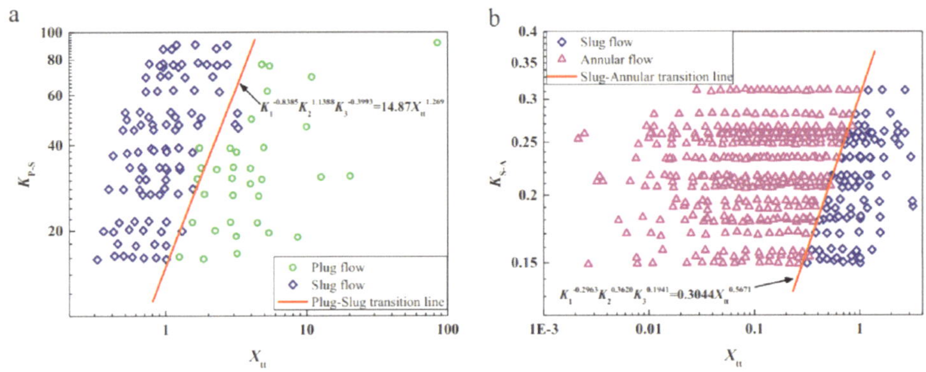

Figure 5.
Flow pattern map of Yang et al. [5].

Dimensionless number K1 is a ratio of evaporation momentum force with inertia force:

$$K_1 = \frac{\text{evaporation momentum force}}{\text{inertia force}} = \frac{\left(\frac{q}{h_{lv}}\right)^2 \frac{1}{\rho_v}}{\frac{G^2}{\rho_l}} = \left(\frac{q}{h_{lv}}\right)^2 \frac{\rho_l}{\rho_v} \tag{21}$$

Dimensionless number K2 is a ratio of evaporation momentum force with surface tension force:

$$K_2 = \frac{\text{evaporation momentum force}}{\text{surface tension force}} = \frac{\left(\frac{q}{h_{lv}}\right)^2 \frac{1}{\rho_v}}{\frac{\sigma}{D}} = \left(\frac{q}{h_{lv}}\right)^2 \frac{D}{\sigma\rho_v} \tag{22}$$

Dimensionless number K3 is a ratio of shear force with gravity force:

$$K_3 = \frac{\text{shear force}}{\text{gravity force}} = \frac{\frac{\mu_l}{\rho_l}\frac{G}{D}}{(\rho_l - \rho_v)gD} = \frac{\mu_l G}{(\rho_l - \rho_v)\rho_l g D^2} \tag{23}$$

The transition line from plug to slug and slug to annular based on function of Xtt:

Plug to slug:

$$K = K_{P-S} = K_1^{-0.8385} K_2^{1.1388} K_3^{-0.3993} = 14.87 X_{tt}^{1.269} \tag{24}$$

Slug to annular:

$$K = K_{S-A} = K_1^{-0.2963} K_2^{0.3620} K_3^{0.1941} = 0.3044 X_{tt}^{0.5671} \tag{25}$$

3. Two phases heat transfer coefficient in boiling

Due to the changing phase during the convective boiling, the proportion of liquid and vapor is also changed, which affected the mechanism of boiling. In order to indicate the boiling heat transfer coefficient, the nucleate boiling and convective boiling are considered. More detailed about the methodology, the boiling number and some other factors are presented in the method of each group authors as below:

3.1 The correlation of Han et al.

Base on the basic form of boiling heat transfer coefficient, Han et al. [20] was established their correlation from the updating their experimental result. The experimental data results were done on the working range of mass flux G = [100; 250] kg/m^{2}s, heat flux q = [11.76; 52.94] kW/m^2, temperature T = [−5, 8]°C with fluid R161 for microfin tube with 6.34 mm average inside diameter, 15° helix angle, 30° fin angle, 0.1 mm fin height, 65 number of fins.

Heat transfer coefficient of two-phase flow:

$$h_{r,tp} = Fh_{r,l} + Sh_{r,nb} \tag{26}$$

Convective heat transfer with convective fin factor:

$$h_{r,l} = E_{RB}h_l \tag{27}$$

$$h_l = 0.023\,Re_l^{0.8}Fr_l^{0.4}\left(\frac{k_l}{d_i}\right) \tag{28}$$

$$E_{RB} = \left\{1 + \left[2.64(Re_l)^{0.036}\left(\frac{e}{d_i}\right)^{0.212}\left(\frac{p}{d_i}\right)^{-0.21}\left(\frac{\beta}{90^o}\right)^{0.29}(Pr_L)^{-0.024}\right]^7\right\}^{1/7} \tag{29}$$

$$Re_{rl} = \frac{G(1-x)d_i}{\mu_l} \tag{30}$$

Heat transfer of nuclear boiling

$$h_{r,nb} = 55P_r^{0.12}(-\log_{10}P_r)^{-0.55}M^{-0.5}q^{0.67} \tag{31}$$

The new function of F and S

$$F = 1 + aBo^{1.16} + b\left(\frac{1}{X_{tt}}\right)^{0.86} \tag{32}$$

$$S = \frac{1}{1 + cF^d\,Re_l^{1.17}} \tag{33}$$

In which a = 7196.741; b = 1.5135; c = 2.703; d = 1.94,

where e: microfin height; p: axial pitch from fin to fin; β: helix angle; N: number of fin; Pr: reduced pressure.

3.2 The correlation of Rollmann and Spindler

The model of Rollmann and Spindler [13] for heat transfer coefficient was derived with total 1614 data points experiment for refrigerant R407C on microfin tube at 8.95 mm fin root diameter, 0.24 mm fin height, 15° Helix angle, 25° Apex angle and 55 fins. Heat flux q = [1000; 20000] W/m^2, mass flux G = [25; 300] kg/m^{2}s, saturation temperature Tsat = [−30; 10]°C,

The model defined by Nuselt number:

$$Nu(x, Bo, Re, Pr) = C_4\left(\frac{C_1}{Pr^2} + C_2\right)Re^{2/3}[\ln(Bo) + C_3]x^{\left(\frac{C_1}{Pr^2}+C_2\right)} \tag{34}$$

where
$C_1 = -3.7$; $C_2 = 0.71$; $C_3 = 12.17$; $C_4 = 1.2$
$Bo = \frac{q}{G\Delta h_v}$; $Re = \frac{GD_{FR}}{\eta_L}$; $Pr = \frac{\eta_L C_{p,L}}{\lambda_L}$; $Nu = \frac{hD_{FR}}{\lambda_L}$
DFR: diameter at fin root
Working range: Bo >5.1837.10−6; Pr > 2.2828

3.3 Correlation for boiling model of Chamra and Mago

The semi empirical model of Chamra and Mago [11] was derived based on 380 collected data points from available literature.

Working range of collected database	The range of microfin tube on database
Fluid: R134A, R12, R22, R123,	d_i = [7.92; 15.88] mm (inner tube diameter)
Ts = [0.6; 15]°C	n_f = [21; 100] number of microfin per unit length
q = [0; 64.3] kW/m^2	e = [0.12; 0.38] mm (fin height)
x = [0.05; 1]	β = [10; 90]° (apex angle)
G = [25; 410] kg/m^{2}s	γ = [3; 30]° (helix angle)
	L = [0.3-4.88] m
	t_h = [0.28-0.51] mm (tube wall thickness)

Developed on the basic of Cavallini model

$$h_{tp} = h_{pb}1.5160(X_{tt})^{1.1610}F_1^{-1.7640} + h_l\Phi Rx^{2.6220}\left(Bon^w Fr_V\right)^{-0.2158}F_2^{0.5927}F_3^{0.0582} \tag{35}$$

$$h_{pb} = 55P_R^{0.12}\left(-\log_{10}P_R\right)^{-0.55}M^{-0.5}q^{0.67} \tag{36}$$

$$q = G.i_{fg}.\Delta x \tag{37}$$

$$X_{tt} = \left(\frac{1-x}{x}\right)^{0.9}\left(\frac{\rho_v}{\rho_l}\right)^{0.5}\left(\frac{\mu_l}{\mu_v}\right)^{0.1} \text{ If } X_{tt} > 1 \text{ then } X_{tt} = 1 \tag{38}$$

$$F_1 = 0.01d_i^{-1} \tag{39}$$

$$h_l = 0.023\frac{k_l}{d_i}Re_l^{0.8}Pr_l^{0.4} \tag{40}$$

$$\Phi = \left[(1-x) + 2.63x\left(\frac{\rho_l}{\rho_v}\right)^{\frac{1}{2}}\right]^{0.8} \tag{41}$$

$$Rx = \left\{\frac{2en_f(1-\sin(\beta/2))}{\pi d_i\cos(\beta/2)} + 1\right\}\frac{1}{\cos(\gamma)} \tag{42}$$

$$Bon^w = \frac{g\rho_L e\pi d_i}{8\sigma n_f} \tag{43}$$

$$Fr_V = \frac{G^2}{\rho_v^2 g d_i} \tag{44}$$

$$F_2 = 0.01d_i^{-1} \tag{45}$$

$$F_3 = 100G^{-1} \tag{46}$$

3.4 The correlation of Yun et al.

Using the database with 749 data points of five different refrigerants to create a generalized correlation for boiling heat transfer in horizontal microfin tubes.

Heat transfer coefficient of two-phase flow:

$$h_{tp}/h_l = \left[C_1 Bo^{C_2}\left(\frac{P_{sat}d_i}{\sigma}\right)^{C_3} + C_4\left(\frac{1}{X_{tt}}\right)^{C_5}\left(\frac{Gf}{\mu_l}\right)^{C_6}\right] Re_l^{C_7} Pr_l^{C_8}\left(\frac{\delta}{f}\right)^{C_9} \tag{47}$$

$$Bo = \frac{q}{Gh_{lv}} \tag{48}$$

$$h_l = 0.023\, Re_l^{0.8} Fr_l^{0.4}\left(\frac{k_l}{d_i}\right) \tag{49}$$

$$Re_l = \frac{G(1-x)d_i}{\mu_l} \tag{50}$$

$$\delta = \frac{d_r(1-\varepsilon)}{4} \tag{51}$$

$$\varepsilon = \frac{x}{\rho_G}\left[(1+0.12(1-x))\left(\frac{x}{\rho_G}+\frac{1-x}{\rho_L}\right)+\frac{1.18(1-x)[g\sigma(\rho_L-\rho_G)]^{0.25}}{G\rho_L^{0.5}}\right]^{-1} \tag{52}$$

$$X_{tt} = \left(\frac{1-x}{x}\right)^{0.9}\left(\frac{\rho_v}{\rho_l}\right)^{0.5}\left(\frac{\mu_l}{\mu_v}\right)^{0.1} \tag{53}$$

Coefficients of correlation	The range of database
C_1 = 0.009622	d = [8.82; 14.66] mm (inner diameter)
C_2 = 0.1106	f = [0.12; 0.381] mm (fin height)
C_3 = 0.3814	spiral angle = [16; 30]°
C_4 = 7.6850	G = [50; 637] kg/m^{2}s
C_5 = 0.5100	q = [5; 39.5] kW/m^2
C_6 = −0.7360	Ts = [−15; 70]°C
C_7 = 0.2045	R22, R113, R123, R134A, R410A
C_8 = 0.7452	
C_9 = −0.1302	
d_i: maximum inside diameter of a microfin tube	
d_r: diameter of a microfin tube at fin root	

3.5 The correlation of Cavallini et al.

The correlation of Cavallini et al. [8] produced for not only microfin tube but also for cross groves tubes with 643 data points collected from available literatures.

Heat transfer coefficient of two phase flow:

$$\alpha = \alpha_{nb} + \alpha_{cv} \tag{54}$$

Nucleate boiling component:

$$\alpha_{nb} = \alpha_{cooper} S.F_1(d_i) = \left[55P_R^{0.12}(-\log_{10}P_R)^{-0.55}M^{-0.5}q^{0.67}\right]S.F_1(d_i) \tag{55}$$

$$S = A.X_{tt}^B = A.\left[\left(\frac{1-x}{x}\right)^{0.9}\left(\frac{\rho_G}{\rho_L}\right)^{0.5}\left(\frac{\mu_L}{\mu_G}\right)^{0.1}\right]^B \text{ If } X_{tt} > 1 \text{ then } X_{tt} = 1 \tag{56}$$

$F_1(d_i)$: function of fin tip tube diameter

$$F_1(d_i) = \left(\frac{d_o}{d_i}\right)^C \tag{57}$$

Convective term:

$$\alpha_{cv} = \frac{\lambda_L}{d_i}.Nu_{cv,smooth\ tube}.Rx^S.(Bo.Fr)^T.F_2(d_i).F_3(G) \tag{58}$$

$$Nu_{cv,smooth\ tube} = Nu_{LO}\Phi = \left[0.023\left(\frac{Gd_i}{\mu_L}\right)^{0.8} Pr_L^{1/3}\right]\left[(1-x) + 2.63x\left(\frac{\rho_L}{\rho_G}\right)^{1/2}\right]^{0.8} \tag{59}$$

$$Rx = \left\{\frac{2hn_g(1 - \sin(\gamma/2))}{\pi d_i \cos(\gamma/2)} + 1\right\}\frac{1}{\cos(\beta)} \tag{60}$$

$$Bo = \frac{g\rho_L h\pi d_i}{8\sigma n_g} \tag{61}$$

$$Fr = \frac{u_{GO}^2}{gd_i} \tag{62}$$

$$F_2(d_i) = \left(\frac{d_o}{d_i}\right)^V \tag{63}$$

$$F_3(G) = \left(\frac{G_o}{G}\right)^Z \tag{64}$$

u_{GO}: velocity of gas phase with total flow rate

	A	B	C	S	T	V	Z	do	Go
$G < 500$ kg/m²s	1.36	0.36	0.38	2.14	−0.15	0.59	0.36	0.01	100
$G \geq 500$ kg/m²s	1.36	0.36	0.38	2.14	−0.21	0.59	0.36	0.01	100

Working range:

Geometry of tubes:	Experimental Data bank: 643
d_i = [3; 14.3] mm (minimum inside tube diameter)	Fluid: R134A, R12, R22, R123, R125, R32
ng = [30; 112] (number of grooves)	Ts = [−6.6; 48]°C
h = [0.1; 0.35] mm (fin height)	q = [3; 82] kW/m²
γ = [20; 120]° (apex angle)	x = [0.05; 0.9]
β = [4; 30]° (spiral angle)	G = [90; 600] kg/m²s
L = [0.2; 3.67] m	

3.6 The correlation of Thome et al.

The model was derived to predict the microfin with the test data for R134a, R123, mass flux G = [100; 500] kg/m²s, quality x = [0.15; 0.85] and heat flux q = [2; 47] kW/m²

$$h = E_{mf}\left[(h_{nb})^3 + (E_{RB}h_{cv})^3\right]^{1/3} \tag{65}$$

$$h_{nb} = 55P_R^{0.12}(-\log_{10}P_R)^{-0.55}M^{-0.5}q^{0.67} \tag{66}$$

$$h_{cv} = 0.0133\,Re_l^{0.69}Pr_l^{0.4}\lambda_l/\delta \tag{67}$$

$$(Re_l)_{film} = \frac{4G(1-x)\delta}{(1-\varepsilon)\mu_l} \tag{68}$$

$$\delta = d_r(1-\varepsilon)/4 \tag{69}$$

$$\varepsilon = \frac{x}{\rho_G}\left[(1+0.12(1-x))\left(\frac{x}{\rho_G}+\frac{1-x}{\rho_L}\right)+\frac{1.18(1-x)[g\sigma(\rho_L-\rho_G)]^{0.25}}{G\rho_L^{0.5}}\right]^{-1} \tag{70}$$

$$E_{mf} = 1.89\left(\frac{G}{G_{ref}}\right)^2 - 3.7\left(\frac{G}{G_{ref}}\right) + 3.02 \tag{71}$$

$$E_{RB} = \left\{1+\left[2.64(Re_D)^{0.036}\left(\frac{e}{d_r}\right)^{0.212}\left(\frac{p}{d_r}\right)^{-0.21}\left(\frac{\beta}{90^o}\right)^{0.29}(Pr_L)^{-0.024}\right]^7\right\}^{1/7} \tag{72}$$

$$Re_D = \frac{G(1-x)d_r}{\mu_l} \tag{73}$$

$$p = \frac{\pi d_r/N}{\tan\beta} \tag{74}$$

$$G_{ref} = 500 kg/m^2s \tag{75}$$

where e: microfin height; p: axial pitch from fin to fin; β: helix angle; N: number of fins; dr: root diameter

4. Pressure drop

4.1 The correlation of Choi et al.

The model form of Choi et al. [14] was carryout from Pierre 1964 model with 831 data pointed collected from NIST database with 626 data point for boiling and 205 data points for condensation for some different fluids R134a, R22, R125, R32, R407C, R410A and R32/R134a. Those data points derived with test section has 8.92 mm root diameter, 9.52 mm outside diameter, 18° helix angle.

Total pressure drop:

$$\frac{\Delta p}{L} = \frac{\Delta p_f}{L} + \frac{\Delta p_m}{L} = G^2\left[f\frac{(v_{tp,out}+v_{tp,in})}{d_h} + \frac{(v_{tp,out}-v_{tp,in})}{L}\right] \tag{76}$$

Two phase friction factor:

$$f = 0.00506\,Re_{h,LO}^{-0.0951}K_f^{0.1554} \tag{77}$$

$$Re_{h,LO} = Gd_h/\mu_l \tag{78}$$

Hydraulic diameter:

$$d_h = 4A_c\cos\beta/(n.S_p) \tag{79}$$

Two phase number:

$$K_f = \frac{\Delta x.h_{lv}}{gL} \tag{80}$$

"Specific volumes of the two-phase fluid, $v_{tp,out}$ and $v_{tp,in}$, are quality-weighted sums of the vapor and liquid specific volumes at either the outlet or inlet of the tube"

4.2 The correlation of Goto et al.

Model of Goto et al. [15] derived from experiment of R41A and R22 for both boiling and condensation at mass flux G = [200; 340] kg/m^{2}s inside spiral groove tube 7.3 mm mean inside diameter and herring-born groove tube 7.24 mm mean inside diameter.

Frictional pressure gradient:

$$\left(\frac{dP}{dz}\right)_f = \Phi_v^2\left(\frac{dP}{dz}\right)_v = \Phi_v^2 2f_{e,v,Go}(Gx)^2/(\rho_v d_e) \tag{81}$$

$$\Phi_v = 1 + 1.64X_{tt}^{0.79} \tag{82}$$

Or could be compute based on the liquid phase form

$$\left(\frac{dP}{dz}\right)_f = \Phi_l^2\left(\frac{dP}{dz}\right)_l = \Phi_l^2 2f_{e,l,Lo}(G(1-x))^2/(\rho_l d_e) \tag{83}$$

$$\Phi_l = 1 + 7.61X_{tt}^{-1.7} \tag{84}$$

$$Re_{e,v} = Gxd_e/\mu_v \text{ Diameter of inner tube de} \tag{85}$$

$$X_{tt} = \left(\frac{1-x}{x}\right)^{0.9}\left(\frac{\rho_v}{\rho_l}\right)^{0.5}\left(\frac{\mu_l}{\mu_v}\right)^{0.1} \tag{86}$$

$$Re_{e,v} \le 2000; => f_{e,v,Go} = 16/Re_{e,v} \tag{87}$$

$$2000 \le Re_{e,v} \le 2600; => f_{e,v,Go} = 0.000147\, Re_{e,v}^{0.53} \tag{88}$$

$$2600 \le Re_{e,v} \le 6500; => f_{e,v,Go} = 0.046\, Re_{e,v}^{-0.2} \tag{89}$$

$$6500 \le Re_{e,v} \le 12700; => f_{e,v,Go} = 0.00123\, Re_{e,v}^{0.21} \tag{90}$$

$$12700 \le Re_{e,v}; => f_{e,v,Go} = 0.0092 \tag{91}$$

4.3 The correlation of Wongsa et al.

The correlation for boiling with high mass flux was proposed by Wongsa et al. [16]. Derived from their experimental results for R134a on microfin tube with 8.92 mm inner diameter, saturation temperature Tsat = [10; 20]°C, mass flux G = [400; 800] kg/m^{2}s.

Two phase friction pressure drop

$$\left(\frac{dP}{dz}\right)_f = \Phi_{VO}^2\left(\frac{dP}{dz}\right)_{h,VO} = \Phi_{VO}^2 2f_{h,VO}G^2/(\rho_v d_h) \tag{92}$$

$$4f_{h,VO} = \frac{1.325}{\left[\ln\left(\frac{Rx_h}{3.7} + \frac{5.74}{Re_{h,VO}^{0.9}}\right)\right]^2} \tag{93}$$

$$Re_{h,VO} = \frac{Gd_h}{\mu_v} \tag{94}$$

$$Rx_h = \frac{0.3(e/d_h)}{(0.1 + \cos\beta)} \tag{95}$$

$$\Phi_{VO}^2 = 2.3263 - 1.8043\left\{\frac{X_{tt}G}{[gd_h\rho_v(\rho_l - \rho_v)]^{0.5}}\right\}^{0.0802} \tag{96}$$

4.4 The correlation of Kou and Wang

The correlation of Kou and Wang 1996 [17] was analyzed from their own experimental data results for boiling of R22 and R407C in a 9.52 mm diameter micro-fin tube and a smooth tube at two different evaporation temperatures (6°C and 10°C). The mass flux was between 100 and 300 kg/m^{2}s and the heat fluxes 6 and 14 kW/m^2.

$$\left(\frac{dP}{dz}\right)_f = \frac{0.0254G^2[xv_v + (1 - x)v_l]}{d_{fr}} \tag{97}$$

5. Example

Determine the heat transfer coefficient and pressure drop of refrigerant R1234ze during boiling process inside horizontal microfin tube at 5°C and heat flux 8,62 kW/m^2, mass flux 222 kg/m^{2}s. The geometry of microfin tube is 60 number of fins, the inner surface diameter is 8.96 mm, fin height is 0.2 mm, the helix angle and apex angle are 18° and 40°, respectively.

Solve:

At saturation temperature t_{sat} = 5°C of refrigerant R1234ze, thermal properties could be taken in **Table 1**.

Refrigerant: R1234ze, molar mass: M = 114 g/mol, $p_{critical}$ = 3.64 MPa

Reduce pressure: $p_r = p_{sat}/p_{critical}$ = 0.2593/3.64 = 0.07124

Heat flux: q = 8.62 kW/m^2

Mass flux: G = 222 kg/m^{2}s

Microfin tube: N = 60 number of fins, dr = 8.96 mm, e = 0.2 mm, β = 18o; γ = 40°

Property	Unit	Value	Property	Unit	Value
T_{sat}	[°C]	5	P_{sat}	[MPa]	0.2593
ρ_L	[kg/m^3]	1225.5	λ_L	[W/m-K]	8.14E-02
ρ_V	[kg/m^3]	13.9	λ_V	[W/m-K]	1.20E-02
ν_L	[m^3/kg]	8.16E-04	μ_L	[Pa-s]	2.53E-04
ν_v	[m^3/kg]	7.18E-02	μ_V	[Pa-s]	1.14E-05
hL_V	[kJ/kg]	181	Pr_L	[−]	4.102
Cp_L	[kJ/kg-K]	1.319	Pr_V	[−]	0.86
Cp_V	[kJ/kg-K]	0.898	σ	[N/m]	1.15E-02

Table 1.
Properties of R1234ze.

5.1 Flow pattern map

Apply the flow pattern map of Rollmann and Spindler [1] to present in here. At first, calculate the transition lines between one regime to another as a function of quality x and use the classification of the regimes on the map:

Conditions	Regime
$G < G_{strat}$	Stratified flow
$G > G_{slug}$ and $G < G_{sw}$	Stratified wavy flow
$G < G_{slug}$ and $G < G_{sw}$	Stratified wavy/slug flow or slug flow
$G < G_{slug\text{-}helix}$	Slug/helix flow
$G < G_{helix}$	Helix flow
$G > G_{helix}$	Annular flow

In this case, the flow pattern map for boiling of R1234ze in microfin tube was built at 5 °C saturation temperature, heat flux q = 8.62 kW/m^2, with the fixed mass flux G = 222 kg/m^{2}s.

The result map shown in **Figure 6**, when the mass flux lower than 50 kg/m^{2}s, the flow boiling during the changing phase just only occur at the fully stratified flow. But if working condition at mass flux 200 kg/m^{2}s, at the beginning of boiling process, the quality is still low, the boiling occur at slug/helix flow until the quality reach to 0.15, the helix flow happen, and keep boiling with the helix flow until quality is 0.25, it moves to the annular flow boiling to the rest of changing phase.

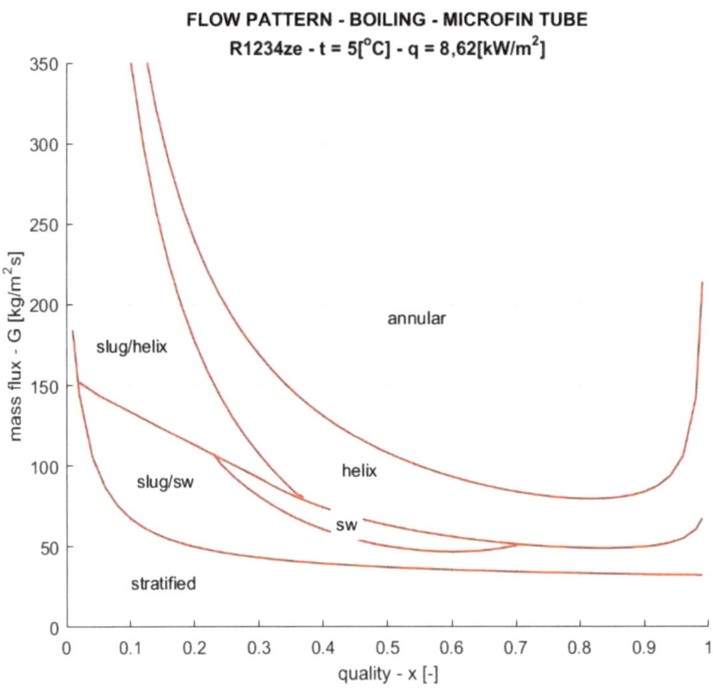

Figure 6.
Flow pattern map on microfin tube.

5.2 Heat transfer coefficients

Could take quality x = 0.5 as an example to present the procedure to calculate heat transfer coefficient and pressure drop of boiling refrigerant inside horizontal microfin tube, with the data point of quality x from 0 to 1 can be determined with the same method.

5.2.1 Applied the correlation of Thome et al.

Apply equation from (65) to (75) to calculate heat transfer coefficients as below:
Void fraction

$$\varepsilon = \frac{x}{\rho_G}\left[(1+0.12(1-x))\left(\frac{x}{\rho_G}+\frac{1-x}{\rho_L}\right)+\frac{1.18(1-x)[g\sigma(\rho_L-\rho_G)]^{0.25}}{G\rho_L^{0.5}}\right]^{-1}$$

$$\varepsilon = \frac{0.5}{13.9}\left[(1+0.12(1-0.5))\left(\frac{0.5}{13.9}+\frac{1-0.5}{1225.5}\right)+\frac{1.18(1-0.5)[9.81x0.0115(1225.5-13.9)]^{0.25}}{222x1225.5^{0.5}}\right]^{-1}$$

$$\varepsilon = 0.9266$$

$$\delta = \frac{d_r(1-\varepsilon)}{4} = \frac{8.96.10^{-3}(1-0.9266)}{4} = 0.00016m$$

$$Re_l = \frac{4G(1-x)\delta}{(1-\varepsilon)\mu_l} = \frac{4x222x(1-0.5)x0.00016}{(1-0.9266)x0.000253} = 3931.067$$

$$h_{cv} = \frac{0.0133\, Re_l^{0.69} Pr_l^{0.4}\lambda_l}{\delta} = \frac{0.0133(3931.067^{0.69})(4.102^{0.4})0.0814}{0.00016} = 3497.97$$

$$\begin{aligned} h_{nb} &= 55P_R^{0.12}(-\log_{10}P_R)^{-0.55}M^{-0.5}q^{0.67} \\ &= 55x0.07124^{0.12}(-\log_{10}0.07124)^{-0.55}114^{-0.5}8620^{0.67} = 1507.309 \end{aligned}$$

$$G_{ref} = 500kg/m^2s$$

$$E_{mf} = 1.89\left(\frac{G}{G_{ref}}\right)^2 - 3.7\left(\frac{G}{G_{ref}}\right) + 3.02 = 1.89\left(\frac{222}{500}\right)^2 - 3.7\left(\frac{222}{500}\right) + 3.02 = 1.75$$

$$Re_D = \frac{G(1-x)d_r}{\mu_l} = \frac{222(1-0.5)0.00896}{0.000253} = 3931.07$$

$$p = \frac{\pi d_r/N}{\tan\beta} = \frac{\pi x0.00896/60}{\tan(18^o)} = 0.001444$$

$$E_{RB} = \left\{1+\left[2.64(Re_D)^{0.036}\left(\frac{e}{d_r}\right)^{0.212}\left(\frac{p}{d_r}\right)^{-0.21}\left(\frac{\beta}{90^o}\right)^{0.29}(Pr_L)^{-0.024}\right]^7\right\}^{1/7} = 1.43$$

$$E_{RB} = \left\{1+\left[2.64(3931.07)^{0.036}\left(\frac{0.0002}{0.00896}\right)^{0.212}\left(\frac{0.001444}{0.00896}\right)^{-0.21}\left(\frac{18^o}{90^o}\right)^{0.29}(4.102)^{-0.024}\right]^7\right\}^{1/7}$$

$$E_{RB} = 1.43$$

$$\begin{aligned} h &= E_{mf}\left[(h_{nb})^3+(E_{RB}h_{cv})^3\right]^{1/3} = 1.75\left[(1507.309)^3+(1.43x3497.97)^3\right]^{1/3} \\ &= 8831.04\left(\frac{W}{m^2K}\right) \end{aligned}$$

5.3 Pressure drop

5.3.1 The correlation of Goto et al.

Apply equation from (81) to (91) to calculate pressure drop as below:
Frictional pressure gradient:

$$\left(\frac{dP}{dz}\right)_f = \Phi_v^2\left(\frac{dP}{dz}\right)_v = \Phi_v^2 2f_{e,v,Go}(Gx)^2/(\rho_v d_e)$$

$$X_{tt} = \left(\frac{1-x}{x}\right)^{0.9}\left(\frac{\rho_v}{\rho_l}\right)^{0.5}\left(\frac{\mu_l}{\mu_v}\right)^{0.1} = \left(\frac{1-0.5}{0.5}\right)^{0.9}\left(\frac{13.9}{1225.5}\right)^{0.5}\left(\frac{0.000253}{0.0000114}\right)^{0.1}$$
$$= 0.1452$$

$$\Phi_v = 1 + 1.64X_{tt}^{0.79} = 1 + 1.640x1452^{0.79} = 1.3571$$

$$Re_{e,v} = \frac{G.x.d_e}{\mu_v} = \frac{222x0.5x0.00896}{0.0000114} = 87242.11$$

Compare $Re_{e,v}$ to get $f_{e,v,Go}$

$$Re_{e,v} \le 2000; \; => f_{e,v,Go} = 16/Re_{e,v}$$

$$2000 \le Re_{e,v} \le 2600; \; => f_{e,v,Go} = 0.000147\,Re_{e,v}^{0.53}$$

x [−]	**h [W/m2s]**	**dP/dz. [Pa/m]**
0.01	2807.2	78.0
0.05	3640.1	248.5
0.10	4712.6	532.1
0.15	5561.5	811.6
0.20	6234.6	1115.3
0.25	6789.2	1441.7
0.30	7265.0	1789.0
0.35	7689.4	2155.5
0.40	8082.1	2539.5
0.45	8458.3	2939.1
0.50	8831.0	3352.5
0.55	9212.8	3777.6
0.60	9617.1	4212.3
0.65	10060.3	4653.8
0.70	10564.8	5099.2
0.75	11164.6	5544.6
0.80	11917.8	5984.8
0.85	12938.2	6412.2
0.90	14500.7	6814.0
0.95	17596.5	7162.5
0.99	27601.3	7331.9

Table 2.
Heat transfer coefficient and pressure drop during the boiling process.

$$2600 \le \text{Re}_{e,v} \le 6500; \; = > f_{e,v,Go} = 0.046\,\text{Re}_{e,v}^{-0.2}$$

$$6500 \le \text{Re}_{e,v} \le 12700; \; = > f_{e,v,Go} = 0.00123\,\text{Re}_{e,v}^{0.21}$$

$$12700 \le \text{Re}_{e,v}; \; = > f_{e,v,Go} = 0.0092$$

With $12700 \le \text{Re}_{e,v} = 87242.11; \; = > f_{e,v,Go} = 0.0092$

$$\left(\frac{dP}{dz}\right)_f = \Phi_v^2\left(\frac{dP}{dz}\right)_v = \frac{\Phi_v^2 2f_{e,v,Go}(Gx)^2}{(\rho_v d_e)} = \frac{(1.3571)^2 x2x(0.0092)(222x0.5)^2}{(13.9x0.00896)}$$
$$= 3352.52\left(\frac{Pa}{m}\right)$$

Apply the same procedure at each data point of quality x of refrigerant from 0.1 to 0.99 to determine heat transfer and pressure drop during the convective boiling. Data results are obtained in **Table 2**.

6. Conclusions

This chapter is presented the boiling process of refrigerant in horizontal microfin tube. Those are related to the understanding of flow patterns and the procedure to build some flow pattern maps. Some of new correlations to calculate heat transfer coefficient and pressure drop of refrigerant during boiling process have been presented. Detailed step by step to calculate heat transfer performance for typical example is also introduced. However, this is still not enough for this topic, with the new structure of microfin tubes, new refrigerants, this just only the general method to estimate the value, it should be confirmed by experimental method and create the new correlation to extend the wide range of application working conditions.

Acknowledgements

We would like to thank Ho Chi Minh City University of Technology (HCMUT), VNU-HCM for the support of time and facilities for this study.

Author details

Thanh Nhan Phan[1,2]* and Van Hung Tran[1,2]

1 Ho Chi Minh City University of Technology (HCMUT), Ho Chi Minh City, Vietnam

2 Vietnam National University Ho Chi Minh City, Linh Trung Ward, Thu Duc District, Ho Chi Minh City, Vietnam

*Address all correspondence to: phannhan@hcmut.edu.vn

References

[1] P. Rollmann and K. Spindler, "A new flow pattern map for flow boiling in microfin tubes," Int. J. Multiph. Flow, vol. 72, pp. 181–187, 2015.

[2] L. Wojtan, T. Ursenbacher, and J. R. Thome, "Investigation of flow boiling in horizontal tubes: Part I - A new diabatic two-phase flow pattern map," Int. J. Heat Mass Transf., vol. 48, no. 14, pp. 2955–2969, 2005.

[3] D. Biberg, "a Mathematical Model for Two-Phase Stratified Turbulent Duct Flow," Multiph. Sci. Technol., vol. 19, no. 1, pp. 1–48, 2007.

[4] X. Zhuang, M. Gong, G. Chen, X. Zou, and J. Shen, "Two-phase flow pattern map for R170 in a horizontal smooth tube," Int. J. Heat Mass Transf., vol. 102, pp. 1141–1149, 2016.

[5] Z. Q. Yang *et al.*, "A new flow pattern map for flow boiling of R1234ze(E) in a horizontal tube," Int. J. Multiph. Flow, vol. 98, pp. 24–35, 2018.

[6] Thanh Nhan Phan, Van Hung Tran, Nilola Kaloyanov and Momchil Vassilev, "Heat transfer performance of R1234yf for convective boiling in horizontal micro-fin and smooth tubes," E3s Web of Conference 207, vol. 9, 2020.

[7] J. R. Thome, D. Favrat, and N. Kattan, "Evaporation in microfin tubes: A generalized prediction model," *Proc. Convect. Flow Pool Boil. Conf.*, no. May, 1997.

[8] A. Cavallini, D. Del Col, L. Doretti, G. A. Longo, and L. Rossetto, "Refrigerant vaporization inside enhanced tubes: A heat transfer model," Heat Technol., vol. 17, no. 2, pp. 29–36, 1999.

[9] M. Yu, T. Lin, and C. Tseng, "Heat transfer and flow pattern during two-phase flow boiling of R-134a in horizontal smooth and microfin tubes ´ coulement lors de l ' e ´ bullition R134a : transfert de chaleur et e ´ coulement diphasique sur des tubes horizontaux lisses et a en e microa," vol. 25, pp. 789–798, 2002.

[10] R. Yun, Y. Kim, K. Seo, and H. Young Kim, "A generalized correlation for evaporation heat transfer of refrigerants in micro-fin tubes," Int. J. Heat Mass Transf., vol. 45, no. 10, pp. 2003–2010, 2002.

[11] L. M. Chamra and P. J. Mago, "Modelling of evaporation heat transfer of pure refrigerants and refrigerant mixtures in microfin tubes," Proc. Inst. Mech. Eng. Part C J. Mech. Eng. Sci., vol. 221, no. 4, pp. 443–454, 2007.

[12] Z. Wu, Y. Wu, B. Sundén, and W. Li, "Convective vaporization in micro-fin tubes of different geometries," Exp. Therm. Fluid Sci., vol. 44, pp. 398–408, 2013.

[13] P. Rollmann and K. Spindler, "New models for heat transfer and pressure drop during flow boiling of R407C and R410A in a horizontal microfin tube," Int. J. Therm. Sci., vol. 103, pp. 57–66, 2016.

[14] J. Y. Choi, M. a Kedzierski, and P. a Domanski, "Generalized Pressure Drop Correlation for Evaporation and Condensation in Smooth and Micro-Fin Tubes," no. 1 986, 2001.

[15] N. I. M. Goto, N. Inoue, "Condensation and evaporation heat transfer of R410A inside internally grooved horizontal tubes," Int. J. Refrig., vol. 24, no. 1, pp. 528–538, 2001.

[16] J. Wongsa-Ngam, T. Nualboonrueng, and S. Wongwises, "Performance of smooth and micro-fin tubes in high mass flux region of R-134a during evaporation," *Heat Mass Transf. und Stoffuebertragung*, 2004.

[17] C. S. Kuo and R. Laboratories, "In-tube evaporation of HCFC-22 in a 9. 52 mm micro-fin / smooth tube," *Science (80-.).*, vol. 39, no. 12, 1996.

[18] Phan Thanh Nhan, Pressure drop, heat transfer and flow pattern maps inside smooth and enhanced tubes during convective boiling and condensation of R134a and R1234ze [thesis]. Milan, Italy: Politecnical di Milano; 2019.

[19] D. Biberg, "An explicit approximation for the wetted angle in two-phase stratified pipe flow," Can. J. Chem. Eng., vol. 77, no. 6, pp. 1221–1224, 1999.

[20] X. H. Han, Y. B. Fang, M. Wu, X. G. Qiao, and G. M. Chen, "Study on flow boiling heat transfer characteristics of R161/oil mixture inside horizontal micro-fin tube," Int. J. Heat Mass Transf., vol. 104, pp. 276–287, 2017.

Chapter 9

Design, Performance, and Optimization of the Wire and Tube Heat Exchanger

I Made Arsana and Ruri Agung Wahyuono

Abstract

The wire and tube heat exchanger has been mostly utilized as a condenser unit in various refrigeration systems. As a class of extended surface-based heat exchanger, not only the operating condition but also the geometry of the wire and tube heat exchanger plays a critical role in determining the overall performance of the heat exchanger. Despite the fact that the current designs that include the inline, single-staggered, and woven matrix-based wire and tube heat exchangers already exhibits positive performance, future design and optimization remain challenging from the thermal and fluids engineering point of view. To guide the optimization strategy in the heat exchanger design, this chapter provides an insight into how the geometrical design impacts the performance of various wire and tube heat exchangers, which can be deduced from either the heat exchanger capacity or efficiency.

Keywords: heat transfer, extended surface, operating condition, heat exchanger capacity

1. Introduction

The world is recently demanding the energy-efficient technology and process including in industry, building, and urban housing. Among various vital technologies, particularly in building and housing, the heat exchanger is one of the common technologies applied in, for example, air conditioners and refrigeration system. For these applications, it is common that the finned tube heat exchangers have been used [1, 2]. Particularly for the refrigeration system, the wire-based fin has been employed in the so-called wire and tube heat exchanger (or wire-on-tube heat exchanger). This wire and tube heat exchanger consists of a tube coil with certain spacing attached with small diameter wires acting as its extended surface [1–3]. The working fluid, for example, refrigerant, nanofluids, or thermal oil, flows inside the tubes, while the ambient air is exposed across the outside surface of the wire-attached tube coils, which allows for either natural or forced convection to dissipate the heat from the surface [1–5].

There have been many literatures reporting the thermal performances of wire and tube heat exchanger. The seminal work has been reported by Witzell and Fontaine [6, 7] who have investigated the thermal characteristic and the design procedure of wire and tube heat exchanger. Later, heat transfer modes of radiation and natural convection from wire-on-tube heat exchanger have been studied in which the radiation considers all interactions between the surface and the surrounding environment [8]. Following this experimental study, some refrigeration research aiming at formulating the correlation on the air-side heat transfer coefficient in natural convection-based wire and tube heat exchanger has developed, including for a single-layer and a multiple-layer wire and tube [9–12], oscillating heat tubes [13], and wire-woven heat exchanger [14].

Recently, optimizing the wire and tube heat exchanger design into more compact geometry that exhibits high-heat exchanger efficiency and enables reduction the manufacturing cost. As it has been largely used as a condenser in the refrigeration system, reducing the size and material mass while showing high-specific cooling capacity is desirable for optimization. To do so, various numerical thermal models of wire and tube heat exchanger, which are mainly simulated using the finite element methods, have emerged in the last two decades allowing for a comprehensive analysis of the heat transfer process in the heat exchanger [15–19]. The emergence of these numerical studies is indicative of the current research direction on the development of wire and tube heat exchanges. Furthermore, numerical studies will serve as a versatile tool to analyze the performance of the wire and tube-based condenser for various geometrical design parameters, such as wire and tube spacing (pitch), wire and tube diameter, and operating conditions, such as mass flow rate and inlet temperature. It is also important to note that numerical studies will help to reduce the cost of testing and prototyping of modified wire and tube heat exchanger architecture.

In general, in spite of the currently available wire and tube heat exchanger working efficient and reliable, there is always a room for improvement toward optimum heat exchanger design, which is more efficient in terms of thermal efficiency, material mass, and manufacturing cost [16, 17]. It is prevalent that the design optimization using experimental approaches provides an actual figure of the heat exchanger performance. Nonetheless, as mentioned earlier, fabricating geometry-modified wire and tube heat exchanger will cost a considerable amount of money. Therefore, this chapter discusses how the design optimization can be carried out by minimizing effort in the experimental approach and maximizing the use of the experimentally validated numerical model. In this chapter, a custom-built heat exchanger testing apparatus will be presented and used to evaluate the thermal performance of wire and tube heat exchanger. Finite element methods using MATLAB programming and computational fluid dynamics (CFD) approach will be used for the optimization of the geometrical design as well as the operating condition and for understanding the physical phenomena underlying the heat transfer process in the heat exchanger, respectively.

2. Experimental approach

To explore the effect of geometrical design on the wire-tube configuration, three different configurations including inline, single-staggered, and woven matrix wire and tube heat exchanger were fabricated [20]. The design of wire and tube heat exchanger considered various geometrical aspects that include the width (W) of the wire cross, the wire length (Lw), the wire pitch (pw), the wire diameter (Dw), the tube pitch (pt) or tube spacing (s_t), and the tube diameter (Dt). The wire pitch then defined the number of wires used in a certain width of tube coil width (W).
As shown in **Figure 1** (top and side view), the wire-tube connections were different among these three configurations and this difference was expected to affect the heat exchanger performance.

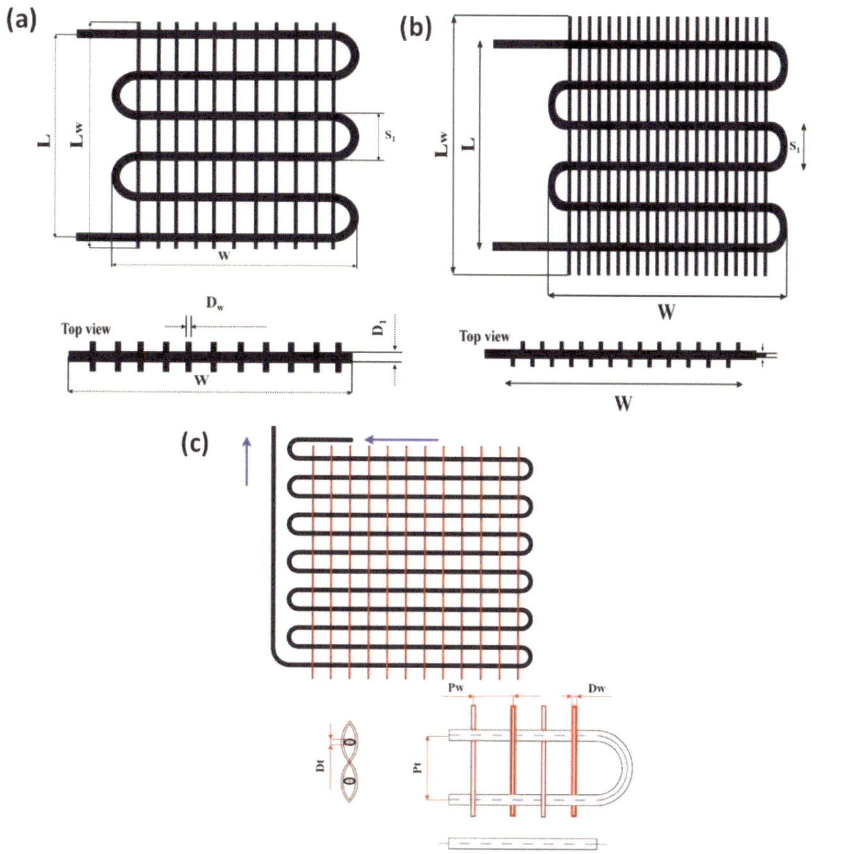

Figure 1.
The wire and tube heat exchanger fabricated and tested in this study with (a) inline, (b) single-staggered, and (c) woven matrix wire-tube configuration. Figure was adapted from Ref. [20] with permission.

The evaluation of heat exchanger capacity and efficiency employed the custom-built heat exchanger testing apparatus as shown in **Figure 2a** [20, 21]. Hot fluid was pumped through the wire and tube heat exchanger, on which the surface temperature was monitored at nine different selected positions. The mass flow rate, pressure, and temperature before and after passing through heat exchangers were monitored as well. Details on the components of the heat exchanger testing apparatus are described in **Figure 2b**, and the specification is summarized in **Table 1**.

2.1 Experimental test for evaluating the wire and tube heat exchanger performance

Running the experiment was started by soaking the working fluid, that is, oil Thermo 22, into the thermostatic vessel. It is worth noting that the apparatus was placed at constant room temperature (T_∞). Afterward, the pump was turned on allowing for the cold fluid to flow in the piping and tubing system. At this juncture, if there was no leakage, the pump was then turned back off. To manipulate the inlet fluid temperature, the working fluid was heated by the installed electric heater inside the vessel, and the temperature was controlled at the desired value. The pump again was turned on to circulate the heated working fluid. In addition, the inlet mass flow could be tuned by adjusting the opening of valves in the apparatus

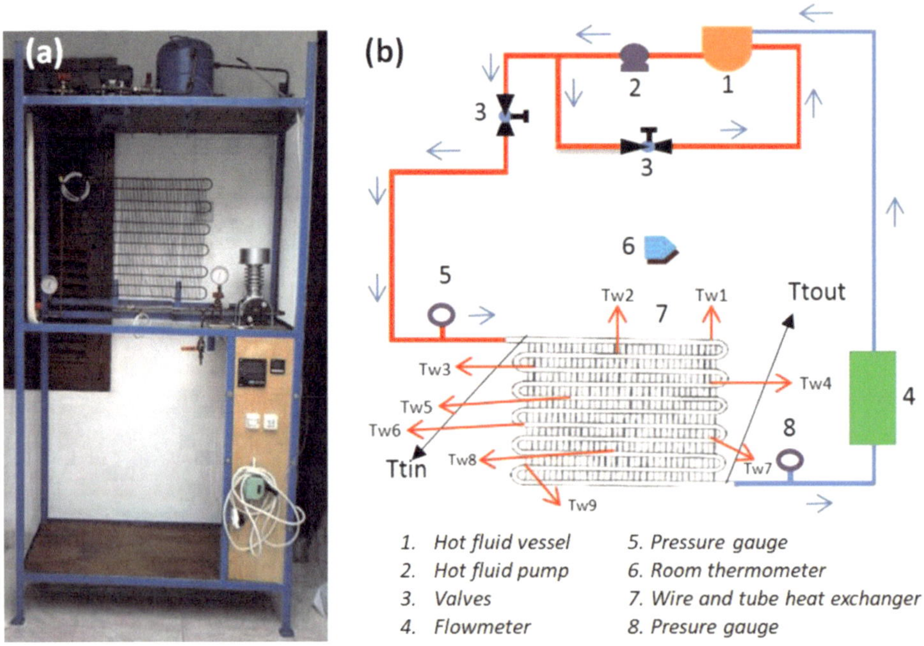

Figure 2.
(a) The visual appearance of the heat exchanger testing apparatus, and (b) schematic of heat exchanger apparatus, which consists of thermostatic vessel embedded with electric heater, pump, pressure gauge, instrumentation box, flow meter, valves, and thermometer. Figure was adapted from Refs. [20,21] with permission.

No.	Component	Specification and operating condition
1	Hot fluid vessel	Atmospheric vessel (1 atm)
2	Hot fluid pump	Trochoid pump (Lamborghini, Italy)
3	Heating System	Thermo-controller: • PXR-9 (Fuji, Japan) • Range: 0–1000°C Thermocouple: • Fluks • Range: 0–200°C Heating element: • Lasco (Germany) • Power: 500 W
4	Flow meter	• Working fluid: oil • Accuracy: 0.1 kg cm^{-3} • Range: 30–100 lpm
5	Pressure gauge	• Zenit & Imperial • Accuracy: 0.2 kg cm^{-2} • Range: 0–5 kg cm^{-2}

Table 1.
The specification and operating conditions of each component in the home-built heat exchanger testing apparatus [20–24].

and checked at the flow meter, which has an accuracy of 0.1 kg cm^{-3}. During the experiment, wire-tube temperatures at nine different points ($T_{w1} - T_{w9}$ shown in **Figure 2b**), the inlet (T_{in}), and outlet (T_{out}) fluid temperature were recorded [22, 23].

2.2 Equations of convection

To evaluate the resulting experimental data, the air properties including density (ρ), kinematic viscosity (υ), Prandtl number (Pr), conduction coefficient (k), thermal diffusivity (α) were interpolated for every T_{out}. These properties were then used to determine the Grashof number (Gr), Rayleigh number (Ra), and Nusselt number (Nu) using the following equation:

$$Gr = \frac{g \cdot \beta \cdot (T_s - T_\infty) \cdot L^3}{v^2} \tag{1}$$

$$Ra = \frac{g \cdot \beta \cdot (T_s - T_\infty) \cdot L^3}{v\alpha} \tag{2}$$

$$Nu = \frac{4}{3}\left(\frac{Gr}{4}\right)^{0.25} \cdot f(\mathrm{Pr}) \tag{3}$$

$$f(Pr) = \frac{0.75\sqrt{Pr}}{\left(0.609 + 1.221\sqrt{Pr} + 1.238Pr\right)^{0.25}} \tag{4}$$

The heat transfer coefficient was then calculated as follows:

$$h = \frac{Nu \cdot k}{L} \tag{5}$$

The heat exchanger capacity (Q) was then calculated based on the heat transfer coefficient in Eq. (5) using the following formula:

$$Q = h \cdot A \cdot (T_s - T_\infty) \tag{6}$$

where A is the total effective area of the wire and tube heat exchanger.

3. Numerical model to evaluate and to optimize the wire and tube heat exchanger performance

Numerical model in this chapter will be discussed based on the two approaches, that is, the finite element method (FEM) whose program was developed and run using a MATLAB program [24, 25] and the FEM using ANSYS Fluent for the computational fluid dynamics (CFD) approach [25–27]. For the FEM developed in MATLAB, the finite element was modeled as wire-tube element which is shown in **Figure 3**. Each modeled element is comprised of a tube whose length is equal to the wire pitch (pw) and the wire, which acts as a fin was set to have a length as long as the tube spacing or pitch (pt). The thermophysical properties of the fluid, which includes mass flow, temperature, enthalpy, and heat, were spatially calculated at the position of x and $x + dx$ for each element, where dx = pw.

The heat transfer in each element from the working fluid inside the tube to the surrounding air followed:

$$Q_{\mathrm{el}} = \mathrm{UA}_{\mathrm{el}}\left(T_f - T_\infty\right)_{\mathrm{el}} \tag{7}$$

where the conductance variable $\mathrm{UA}_{\mathrm{el}}$ of each element was equal to $\frac{1}{\mathrm{UA}_{\mathrm{el}}} = R_i + R_t + R_o$, and the thermal resistance of each wire-tube element could be expressed:

$$R_{\mathrm{w\&T}} = \left(\frac{1}{h_i A_i} + \frac{ln\,(r_o/r_i)}{2\pi k \Delta z} + \frac{1}{h_o A_o}\right)_{\mathrm{el}} \tag{8}$$

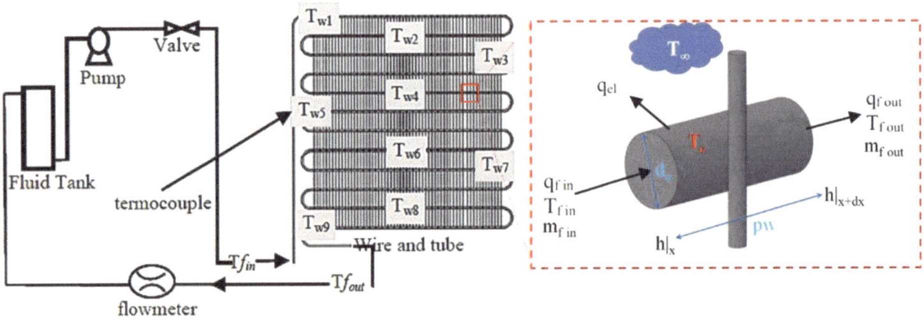

Figure 3.
The schematic of the wire-tube element to build finite element model for wire and tube heat exchanger.

The thermal resistance formula was used as a basis to calculate the convective heat transfer from the fluid to the tube wall, conduction inside the tube wall, and convection from the tube surface to the surrounding air. The area of each element was then determined as $A_o = A_t + A_w = \pi.d_{t,o}.p_w + 2.\pi.d_w.p_t$. As each element of the heat exchanger was extended by a wire-based fin, the wire efficiency could be calculated as follows:

$$\eta_w = \frac{\left[tanh\left(\frac{m\times p_t}{2}\right)\right]}{\left[\left(\frac{m\times p_t}{2}\right)\right]} \tag{9}$$

in which it required convection and conduction heat transfer coefficient data to determine $m = \sqrt{\frac{4h_w}{k_w d_w}}$. For initial calculation of the wire heat transfer coefficient, h_w was set to obtain $\eta_w = 0.9$. Assuming that the heat transfer coefficient is constant along with the wire element and the difference between the fluid temperature and the tube temperature is 0.5°C, the wire temperature was determined as follows:

$$T_w = \eta_w(T_{\text{t,o}} - T_\infty) + T_\infty \tag{10}$$

The average external (outer surface) temperature of each element could then be calculated as follows:

$$T_{\text{ex}} = \frac{(T_{\text{to}} + GP.\eta_w.(T_{t,o} - \text{T}_\infty) + GP.T_\infty)}{(1 + \text{GP})} \tag{11}$$

where GP is the geometrical parameter, $GP = 2\left(\frac{p_t}{d_{\text{t,o}}}\right)\left(\frac{d_w}{p_w}\right)$. The heat transfer coefficient from the outer surface of the wire-tube element was determined from both free convection and radiation $h_o = h_c + h_r$ where the radiative heat transfer coefficient was defined as follows:

$$h_r = \varepsilon.\sigma.\frac{\left(T^4{}_{\text{t,o}} - T^4{}_\infty\right)}{\left(T_{t,o} - \text{T}_\infty\right)} \tag{12}$$

To obtain the convective heat transfer coefficient $h_c = \frac{\text{Nu}\times\text{k}}{H}$, some dimensionless parameters, for example, Nusselt number (Nu) and Rayleigh number (Ra) have to be calculated:

$$Nu = 0.66\left(\frac{Ra.H}{d_{t,o}}\right)^{0.25}\left\{1 - \left[1 - 0.45\left(\frac{d_{t,o}}{H}\right)^{0.25}\right] exp\left(\frac{-s_w}{\phi}\right)\right\} \tag{13}$$

$$Ra = \left(\frac{\beta\rho^2 cp}{\mu k}\right)_a g(T_{t,o} - T_\infty).H^3 \tag{14}$$

$$\phi = \left(\frac{28.2}{H}\right)^{0.4} s_w{}^{0.9} s_t{}^{-1.0} + \left(\frac{28.2}{H}\right)^{0.4} \left[\frac{262}{(T_{t,o} - T_\infty)}\right]^{0.5} s_w{}^{-1.5} s_t{}^{-0.5} p \tag{15}$$

$$s_t = \frac{(p_t - d_t)}{d_t} \tag{16}$$

$$s_w = \frac{(p_w - d_w)}{d_{w.}} \tag{17}$$

Once the h_o was obtained, the ho was compared to h_w. If the difference between h_o and $h_w > 0.1$ W m^{-2} K^{-1}, the ho was set equal to h_w, and the calculation of h_i, h_o, and Q_{el} was run until the convergence criteria were satisfied. At the steady-state condition, the heat transfer from the fluid to the surrounding air was equal to the heat transfer from the fluid to the outer surface of the tube. Thus, the outer surface temperature of the tube, $T_{t,o}$, was determined as follows:

$$T_{t,o} = T_f - Q_{el}\left(\frac{1}{h_i A_i} + \frac{ln\,(r_o/r_i)}{2\pi kl}\right)_{el} \tag{18}$$

The calculated $T_{t,o}$ was then compared to the initial $T_{t,o}$. If the error was more than 0.05°C, new $T_{t,o}$ was substituted into Eq. (11) and the program was run to satisfy the convergence criteria. The calculation resulted in the Q_{el}. For each element, the outlet fluid temperature was determined as follows:

$$T_{out} = \frac{Q_{el}}{\dot{m}} + T_{in} \tag{19}$$

where $\dot{m}$ is the mass flow rate of the working fluid. The output of the i-th element will be used for the calculation for the (i + 1)-th element until the last element of the wire and tube heat exchanger. The summation of all elements yielded Q_{tot} and the heat exchanger capacity was calculated using the following:

$$C = \frac{Q_{tot}}{\dot{m}} \tag{20}$$

The heat exchanger efficiency was determined from the ratio between the actual heat transfer rate to the heat transfer rate if all wire temperature = tube temperature. For heat exchanger temperature equal to the tube temperature, the heat exchanger efficiency followed:

$$\eta_{tot} = \frac{(\eta_w A_w + A_t)}{A_o} \tag{21}$$

For detailed simulation, the heat transfer process that was coupled with the fluid dynamics was modeled and solved using CFD Software package of ANSYS Fluent. Details of the step-by-step procedure of CFD can be found elsewhere [26, 27]. In this simulation, preprocessing, solving, and post-processing steps were respectively executed. The geometry of the wire and tube heat exchanger was modeled in a 1:1 geometric scale and meshed to crease finite elements. Then, the boundary type and boundary condition of modeled geometry were defined. To solve the governing equations (the conservation of energy, mass conservation, and momentum), the initial condition of each boundary condition was inputted. The turbulence model used was Reynold-Average Navier-Stokes (RANS) k-ω SST (shear-stress transport).

This k-ω SST was used since it has a high stability in numerical calculation and could predict accurately the flow on adverse pressure gradient in boundary layer area [25].

4. Performance evaluation of the wire and tube heat exchanger with varying wire-tube configuration

Heat transfer process and its efficiency in the wire and tube heat exchanger are strongly dependent on the geometrical architecture between the wire and tube. In this section, the effect of wire-tube configuration, including inline, staggered, and woven matrix wire and tube, and the operating condition will be discussed based on the experimental investigation and numerical modeling using computational fluiddynamics (CFD).

4.1 The empirical efficiency formulation of the inline wire and tube heat exchanger

To have a starting point for further geometrical optimization, here the simple inline wire and tube heat exchanger is discussed. An experimental investigation using different wire geometries, that is, wire pitch to wire length ratio (pw/Lw), under different operating conditions, that is, inlet fluid temperature of 40–80°C with 10°C intervals, has been carried out [28]. As the heat exchanger has been tested in an isolated and air-conditioned room, the heat exchanger can be assessed employing the relevant parameters and the dimensional analysis to generate a dimensionless equation following the π-Buckingham theory. Thus, the heat exchanger efficiency follows:

$$\eta_o = \frac{q_t}{q_{mox}} = \frac{h.s_w(T_w - T_\infty) + h.s_t(T_t - T_\infty)}{h.s_{tot}(T_t - T_\infty)} \tag{22}$$

$$q_t = f\left[q_{max}, g, \beta, \alpha, \upsilon, \Delta T, pw, dw, Lw, pt, dt\right] \tag{23}$$

The above function can be grouped into several dimensionless parameters: $\pi_1 = \frac{q_1}{q_{max}} = \eta_o$ is the heat exchanger efficiency, $\pi_2 = \frac{g.Lw^3}{\upsilon^2}$ and $\pi_3 = \Delta T.\beta$ are the Grashof number parameters, $\pi_4 = \frac{\alpha}{\upsilon}$ is the reciprocal of Prandtl number, $\pi_5 = \frac{p_w}{L_w}$ is the dimensionless wire pitch, $\pi_6 = \frac{d_w}{L_w}$ is the dimensionless wire diameter, $\pi_7 = \frac{p_t}{L_w}$ is the dimensionless tube pitch, and $\pi_8 = \frac{d_t}{L_w}$ is the dimensionless tube diameter. The equation can then be reformulated as follows:

$$\pi_1 = f\left[\frac{\pi_2.\pi_3}{\pi_4}\pi_5, \pi_6, \pi_7, \pi_8\right] \tag{24}$$

Considering that the wire diameter (dw), the tube diameter (dt), and the tube pitch (pt) are constant, the general mathematical expression of the heat exchanger efficiency can be written:

$$\eta_o = f\left[\frac{g \cdot \beta \cdot \Delta T \cdot Lw^3}{\upsilon \cdot \alpha}, \frac{pw}{Lw}\right] \tag{25}$$

The above function suggests that the heat exchanger efficiency is a function of Rayleigh number (R_a) and the dimensionless wire pitch (or the wire pitch to wire length ratio (pw/Lw)).

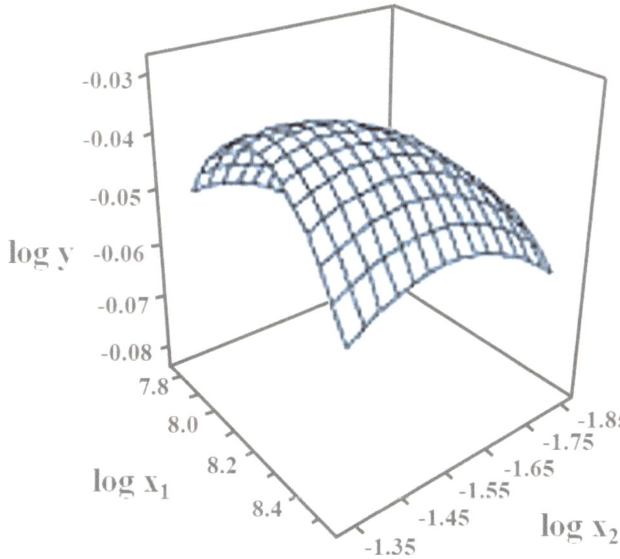

Figure 4.
Contour plot of the heat exchanger efficiency (log y) as a function of Rayleigh number (log x_1) and wire geometry (log x_2, where x_2 = pw/Lw). Figures from Ref. [28] used with permission.

Figure 4 shows the contour plot of the heat exchanger efficiency (y) as function of R_a (x_1) and the wire pitch to wire length ratio (pw/Lw) (x_2). Fitting the experimental data using multivariable logarithmic regression results in the best fitting parameter (R^2 = 0.897) with the resulting function as follows:

$$\begin{aligned} log\,\eta_o &= -8.650 + 2.161\,log\,R_a + 0.259\,log\left(\frac{pw}{lw}\right) - 0.041(\,log\,R_a)^2 \\ &\quad - 0.146\left(\,log\left(\frac{pw}{lw}\right)\right)^2 - 0.084\left(\,log\,R_a \times \,log\left(\frac{pw}{lw}\right)\right) \end{aligned} \tag{26}$$

The above empirical formulation is considered a helpful finding to assist the heat exchanger optimization by taking into account both geometrical aspects of the wire-tube configuration and the operating condition, which is limited by the optimum Ra value ($log^{-1}(x_1)$ at which the log y is maximum).

As above discussed quantitatively, the Ra is dependent on the Prandtl number and Grashof number, and hence, the inlet mass flow inlet as well as the inlet fluid temperature plays role in determining the heat exchange efficiency. To gain its qualitative and quantitative correlation between these operating conditions to the wire and tube heat exchanger efficiency, **Table 2** summarizes the heat exchanger capacity (Q) and heat exchanger efficiency (η) upon varying the inlet mass flow and temperature [22, 23]. In this investigation, the wire and tube heat exchanger possess the following specifications: wire length of 445 mm, wire pitch of 7 mm, wire diameter of 1.2 mm, tube diameter of 5 mm, tube pitch of 476 mm, and tube turn of 12.

At a constant inlet mass flow of 5×10^{-3} kg s^{-1}, the heat exchanger performance improves with increasing inlet fluid temperature. As the temperature increases, the outer surface temperature of wire and tube heat exchanger increases, and hence, the overall heat transfer rate is enhanced, that is, the internal forced convection and conduction from the fluid to the wire-tube elements, and the external radiation from the surface to surrounding and the free convection. Meanwhile, at the same fluid inlet temperature of 70°C increasing the mass flow from 4×10^{-3} kg s^{-1} to

Wire and tube	Controlled parameter	Varied operating condition	Q (W)	η
In line (pw = 7 mm)	M_{in} = 5 × 10^{-3} kg s^{-1}	T_{in} 50°C	9	0.47
		T_{in} 60°C	19	0.65
		T_{in} 70°C	24	0.66
In line (pw = 7 mm)	T_{in} 70°C	m_{in} = 4 × 10^{-3} kg s^{-1}	33.27	0.67
		m_{in} = 5 × 10^{-3} kg s^{-1}	37.62	0.46
		m_{in} = 6 × 10^{-3} kg s^{-1}	41.17	0.73

Table 2.
Heat transfer capacity and heat exchanger efficiency for the inline wire and tube heat exchanger. Data are summarized from Refs. [22,23] upon permission.

6 × 10^{-3} kg s^{-1} enhances the heat exchanger capacity by 24% and the efficiency also increases from 0.67 to 0.73. This increasing heat exchanger capacity is due to the higher mass flow in the tubes the higher the heat transfer rate to the surrounding colder air (at room temperature). This is simply reflected by the increasing inlet to outlet temperature difference ΔT, (T_{in} – T_{out}) which stems from the higher internal forced convection and external free convection as the mass flow increases.

4.2 Inline vs. staggered vs. woven matrix configuration

It is already mentioned that the wire is designed as an extended surface (fin) that is capable to enlarge the heat transfer area in the wire and tube heat exchanger. In this section, this wire-tube configuration will be the focus of the discussion. It is already known that inline and staggered wire and tube heat exchanger have been utilized quite extensively. Nonetheless, some experimentations on the development of woven matrix wire-tube configuration have emerged as it likely combines both the inline and staggered wire-tube configuration.

To compare the heat exchanger capacity among these wire-tube configuration, inline, single-staggered, and woven matrix wire and tube heat exchanger have been fabricated with a variation of wire pitch, that is, 7, 14, and 21 mm [21]. To gain deeper insight, the inlet fluid temperature also varied from 50 to 80°C with 10°C intervals. The heat exchanger apparatus was run at 2 × 10^{-3} kg s^{-1} inlet mass flow and the apparatus was placed in an air-conditioned room at 32°C. **Figure 5** displays the heat exchanger capacity of the wire and tube heat exchanger, which is affected by inlet fluid temperature, wire pitch, and the wire-tube configuration. In general, irrespective of the wire-tube configuration the heat exchanger capacity tends to increase with increasing inlet fluid temperature, whereas the heat exchanger capacity tends to decrease by enlarging the wire pitch irrespective of the wire-tube configuration. For 7 mm wire pitch (**Figure 5a**), inline configuration outperforms other configurations, which is plausible since for inline wire-tube configuration the number of wires used is larger than other configurations. This implies that the extended surface by the surface is enlarged, and thus, the heat transfer rate becomes more efficient.

It is interesting to note that the larger the wire pitch (see **Figure 5b** and **c**) the smaller the difference in heat exchanger capacity among three different wire-tube configurations. Particularly for 21 mm wire pitch, the heat exchanger capacity of a single-staggered configuration is on par with the inline configuration, in which the maximum heat exchanger capacity of 425 W is obtained at 80°C. The heat exchanger efficiency particularly for woven matrix wire and tube heat exchanger is shown in **Figure 5d**. An average of 69% heat exchanger efficiency is obtained

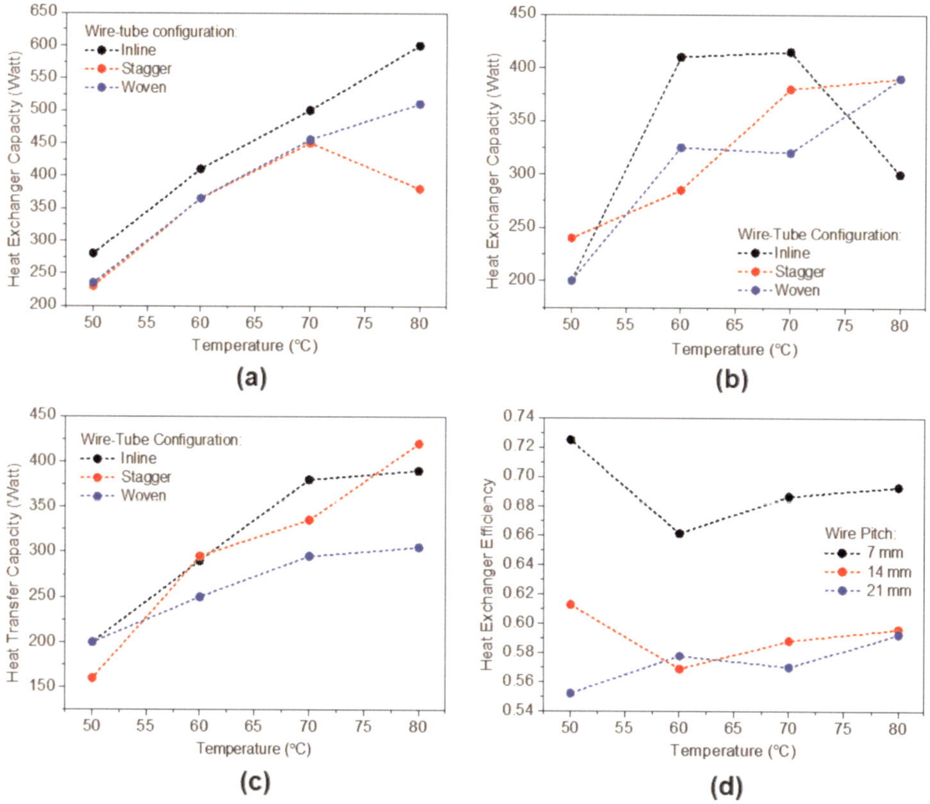

Figure 5.
Different heat exchanger capacity of inline, single-staggered, and woven matrix wire and tube heat exchanger bearing pitch wire of (a) 7 mm, (b) 14 mm, and (c) 21 mm. (d) The efficiency of woven matrix wire and tube heat exchanger using different wire pitches and operated under varying inlet fluid temperature.

for 7 mm wire pitch, while the efficiency drops 10% down by changing the wire pitch to either 14 or 21 mm. This reflects that there is a room for optimization by varying the wire pitch between 7 mm and 14 mm.

Digging into the heat transfer process inside the wire and tube heat exchanger the change in the heat transfer coefficient is shown in **Figure 6** for the utilization of heat exchanger with 7 mm and 14 mm wire pitch. Irrespective of the wire pitch, it is clear that the convection heat transfer coefficient tends to decrease with increasing inlet fluid temperature. While larger wire pitch results in a larger heat transfer coefficient, the inline wire-tube configuration consistently yields the lowest heat transfer coefficient compared to a single-staggered and woven matrix wire-tube configuration. As the convection drops down while the heat exchanger increases at higher inlet fluid temperature, this implies that the conduction from the inner wall to the outer surface of the tube as well as wire tube and the radiation from the surface of the tube to the surrounding dominates the heat transfer process at a higher temperature.

5. Numerical model for heat exchanger performance prediction and optimization

As previously motivated in the introduction, a numerical model of the heat exchanger in general is developed to guide the future optimized design of the heat exchanger. Here, an example of a finite element model has been developed for a

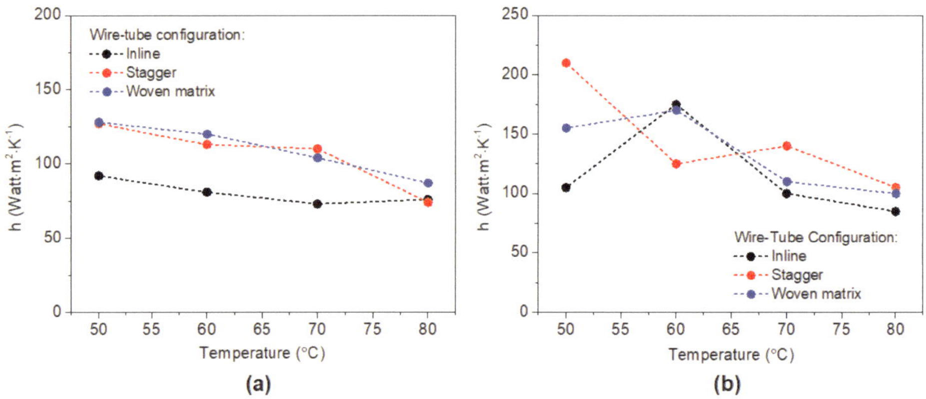

Figure 6.
An exemplified of heat transfer coefficient h *for the inline, single-staggered, and woven matrix wire and tube heat exchanger bearing pitch wire of (a) 7 mm and (b) 14 mm.*

single-staggered wire and tube heat exchanger [24, 25]. In this case, the model was developed to aid the optimum design, which is limited by two variables including wire pitch (pw) and wire diameter (Dw). Experimental validation was carried out for the heat exchanger with wire pitch of 7 mm and 14 mm to assess the accuracy of the finite element model.

Figure 7 highlights the results of the finite element model on the single-staggered wire and tube heat exchanger using a wire pitch of 7 mm. The data verification reveals that the average error of modeling data compared to the measurement data is lower than 5%. In comparison, the earlier finding of the numerical simulation reported by Basal and Chin [16] that employed the heat load validation method yields an average error of up to 10%. This indicates that the simple finite element model developed in this work is considered reliable and accurate.

As the model is experimentally validated and reliable, this model was used to assess the heat exchanger capacity of the wire-tube configuration with the wire pitch spanning from 5 to 12 mm and the wire diameter spanning from 0.8 to 1.5 mm. The contour plot in **Figure 7b** displays the dependency of the heat exchanger capacity (Q_{tot}) to the wire pitch and diameter. For optimization procedure, an optimization factor (f) was defined:

$$f = \frac{Q/W}{Q_0/W_0} \tag{27}$$

where Q is the capacity of the optimized heat exchanger, W is the mass of the heat exchanger, and Q_0 and W_0 are the capacity and mass of the basis designed heat exchanger, respectively.

In this case, the basis designed heat exchanger possess the following specifications: exchanger height of 445 mm; width of heat exchanger (wire) of 431 mm; width of heat exchanger (tube) of 476 mm; the tube length of 6416 mm; the outside tube diameter of 4.8 mm; the inside tube diameter of 3.2 mm; and the wire diameter of 1.2 mm. Using this consideration, the optimum condition is obtained if the optimization factor is maximum. In this regard, the optimum condition is obtained for Q_{tot} of 119.9214 W using a wire pitch of 11 mm and wire diameter of 0.9 mm. The contour plot in **Figure 7b** indicates that the smaller the wire pitch the bigger Q_{tot}. In addition, increasing the wire diameter tends to enhance the Q_{tot}. Nonetheless, if pw ≤ dw, the system will merely be considered as tube, where the fins form of plate structure, and this condition yields lower Q_{tot}.

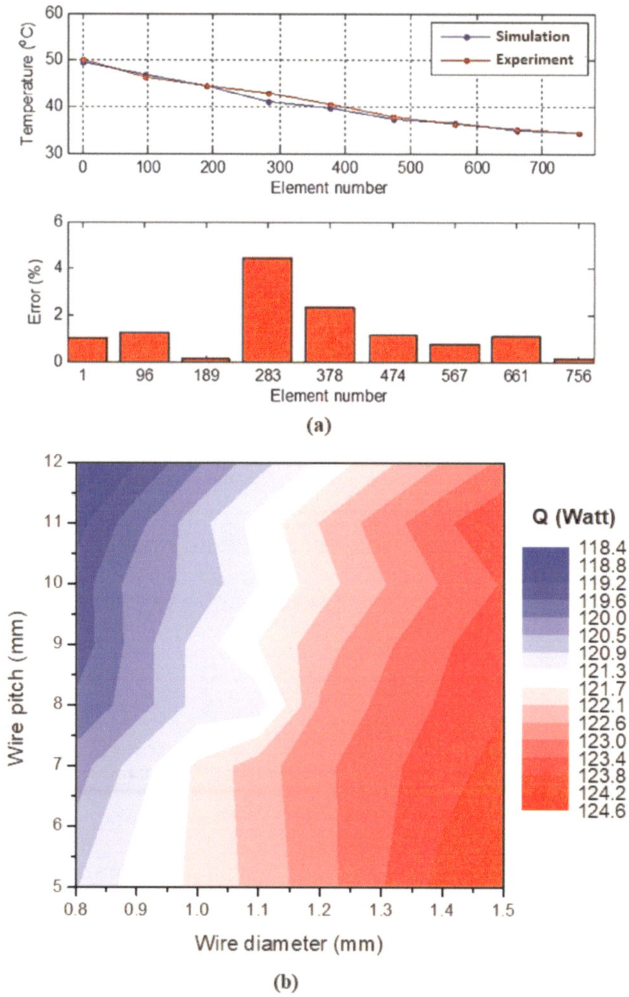

Figure 7.
(a) Temperature validation of each element for heat exchanger with wire pitch of 7 mm representing the wire temperature measured at nine measurement points. (b) Contour plot of the total heat exchanger capacity as a function of wire diameter and wire pitch. Figures from Ref. [24] used with permission.

Having optimized for the design of single-staggered wire and tube heat exchanger, the performance of inline and single-staggered wire-tube configuration is compared experimentally. The experimental data indicate that the inline configuration allows for 110 W heat release, while the single-staggered wire and tube heat exchanger exhibits a heat exchanger capacity of 107.3 W [24]. The difference is not that significant, and this result shows that changing the wire-tube geometry from inline to single-staggered already provides more efficient conditions. It is considered efficient as the single-staggered design reduces the number of wires used for manufacturing, and hence, it cuts the mass of construction materials and costs. Reducing the number of wires also leads to higher convective airflow since the flow is not blocked by the wall geometry as indicated from the velocity contour plot in **Figure 8**.

Figure 9a and **b** displays that air velocity distribution on the array of wire between inline and staggered heat exchanger. The results clearly indicate that the average velocity amplitude of the airflow around the wire tube is higher for a single-staggered one than that of the inline configuration. With velocity amplitude

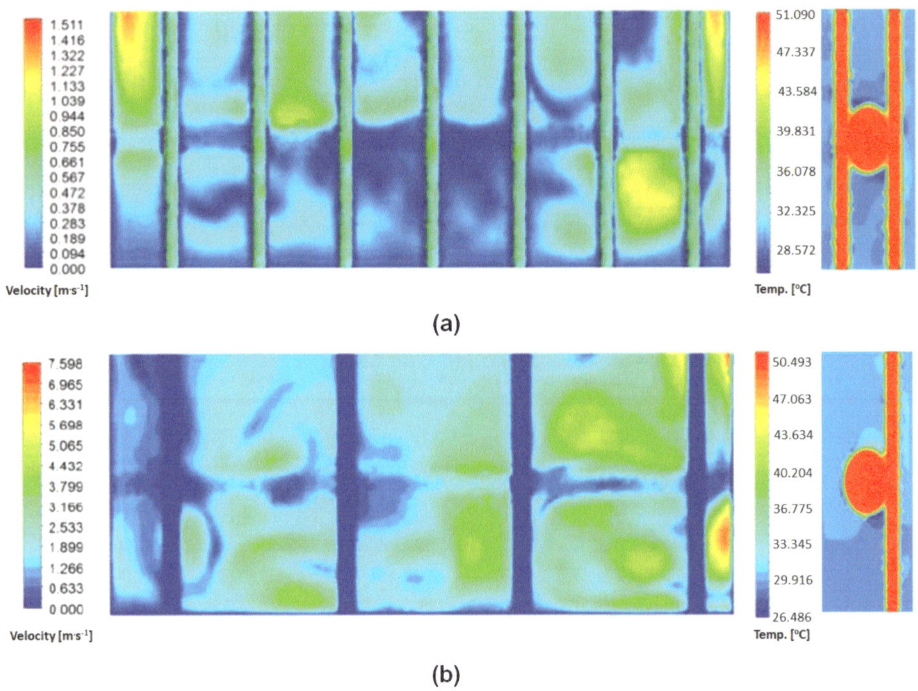

Figure 8.
(Left) Velocity and (right) temperature contour of (a) inline and (b) single-staggered wire and tube heat exchanger bearing wire pitch of 7 mm. Figures from Ref. [24] used with permission.

of ∼4 times higher and more homogeneously distributed, the single-staggered wire-tube heat exchanger also shows a slightly larger temperature difference between the outer surface of the wire-tube and the near-surface air. The more distributed airflow in the vicinity of the staggered wire-tube configuration supports the argument that this configuration minimizes the airflow blocking by geometry wall. As the convective heat transfer coefficient is proportional to the air velocity, this leads to a higher rate of heat release by convection from the surface to the surrounding air.

It is already discussed that the numerical model is helpful to assist the optimization of a single-staggered wire and tube heat exchanger. Here, we further present the numerical study using CFD of single-staggered wire and tube heat exchanger with varying wire pitch (7, 9, and 11 mm) and inlet fluid temperature (40, 60, and 80°C) [26]. In this case, the inlet mass flow was set to 2×10^{-3} kg s^{-1}. The simulation results show that the temperature difference ΔT ($T_{in} - T_{out}$) is exceptionally low and almost similar for the heat exchanger operated in the lowest inlet fluid temperature (**Figure 9a–c**). Thus, the inlet temperature of 40°C yields very low heat exchanger capacity irrespective of the wire pitch. **Figure 9d–f** exhibits quite significant ΔT upon increasing the inlet temperature to 60°C. For wire pitch of 7 mm, a ΔT = 26°C is reached, while wire pitch of 9 and 11 mm shows ΔT of 24 and 22°C, respectively. This simulation is consistent with the experimental results that single-staggered wire and tube heat exchanger bearing 7 mm wire pitch shows ΔT = 26°C [25, 26].

As we have noted previously that heat exchanger capacity tends to increase with increasing inlet fluid temperature, it is also obvious that at 80°C inlet temperature the ΔT becomes significantly higher. The ΔT magnitude for wire pitch of 7, 9, and 11 mm in heat exchanger operated at 80°C inlets are 30, 27, and 25°C, respectively. In brief, the CFD simulation allows for the prediction of the resulting ΔT and the

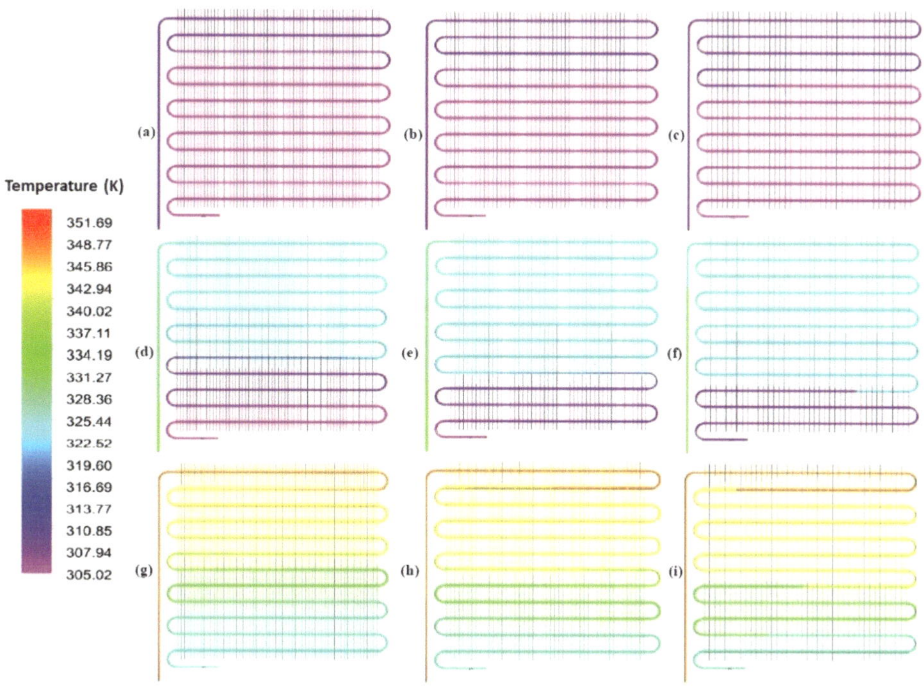

Figure 9.
Temperature profile of the single-staggered wire and tube heat exchanger bearing wire pitch (pw) of (a,d,g) 7 mm, (b,e,h) 9 mm, and (c,f,i) 11 mm operated using inlet fluid temperature of (top) 313 K (40°C), (middle) 333 K (60°C), and (bottom) 353 K (80°C). Figures from Ref. [26] used with permission.

average surface temperature of wire tube that can be used to evaluate heat exchanger capacity, wire efficiency as well as the heat exchanger efficiency.

Another set of examples benefitting the CFD approach for woven matrix wire and tube heat exchanger is shown in **Figure 10**. Previously, it is required for further optimization of woven matrix configuration with wire pitch below 14 mm. Thus, herein we discuss the performance of woven matrix wire and tube heat exchanger bearing wire pitch of 5, 7, and 9 mm and operated at different inlet mass flow, which was simulated using CFD [27]. In this case, the inlet temperature was set to 80°C. The general results show that lowering the mass flow rate from 1.10×10^{-3} to 0.55×10^{-3} kg s^{-1} increases the ΔT. For 5-mm wire pitch, a lower mass flow rate leads to outlet temperature as low as the ambient temperature (29.85°C). Similar results are obtained for heat exchanger bearing 7-mm wire pitch that can yield ΔT = 49.78°C and outlet temperature of 30.4°C under the operating condition of 0.55×10^{-3} kg s^{-1} mass flow rate. Enlarging the wire pitch to 9 mm results in a lower heat transfer capacity as indicated by the higher outlet temperature even though the mass flow rate is set to the lowest. Considering that woven matrix wire and tube heat exchanger bearing wire pitch of 5 and 7 mm have no significant thermal performances, it can be deduced that 7-mm wire pitch is the optimum condition again from both practical and economical aspects.

Using the simulation results in **Figure 10**, the wire efficiency can also be calculated. Overall, increasing the inlet mass flow also improves the wire efficiency. Nonetheless, the maximum efficiency is reached for the inlet mass flow of 0.57×10^{-3} kg s^{-1} irrespective of the wire pitch. The wire pitch in woven matrix configuration affects the wire efficiency, and it is found that 7-mm wire pitch yields the highest efficiency of 73% at the inlet mass flow of 0.57×10^{-3} kg s^{-1} which is consistent with the above discussion.

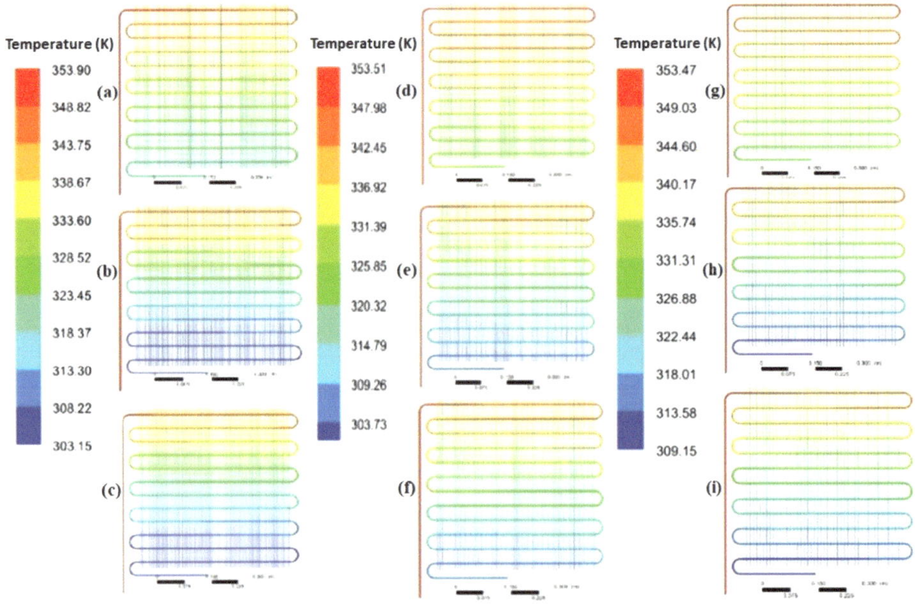

Figure 10.
Temperature profile of single woven matrix wire and tube heat exchanger bearing pitch wire of (a–c) 5 mm, (d–f) 7 mm, and (g–i) 9 mm operated using the inlet mass flow of (top) 1.10×10^{-3} kg s^{-1}, (middle) 0.57×10^{-3} kg s^{-1}, and (bottom) 0.55×10^{-3} kg s^{-1}. Figures from Ref. [27] used with permission.

6. Conclusion

In this chapter, we have discussed that geometrical design along with optimum operating condition substantially control the overall performance of wire and tube heat exchanger. It has been demonstrated that for the commonly used inline wire and tube heat exchanger the efficiency can be predicted using the empirical formula (logarithmic function), which depends merely on the geometrical aspect (wire pitch to wire length ratio) and the Rayleigh number. Consistent with the heat transfer theory in a practical heat exchanger increasing the inlet fluid temperature and slowing down the mass flow rate typically yield higher-heat exchanger capacity and efficiency irrespective of the wire-tube configuration, including the inline, single-staggered, and woven matrix. Nonetheless, among the three investigated wire-tube configurations the single-staggered wire and tube heat exchanger is promising as the most efficient heat exchanger by taking into account the overall thermal performance, material mass, and cost for manufacturing.

This chapter also presents a numerical model based on the finite element method (FEM) that has been developed to evaluate and optimize the geometrical design of a single-staggered wire and tube heat exchanger with the optimization constraints of wire pitch and wire diameter. The FEM has successfully modeled the thermal behavior of the heat exchanger, which is validated by the experimental data showing an average numerical error of less than 5%. To understand the underlying phenomena in the heat transfer process, computational fluid dynamics (CFD) analysis enables the discussion of the local and near-surface airflow affecting the natural convection mechanism, and the heat flux on the outer wire-tube surfaces, which is responsible for the radiation from the surface to the surrounding environment. In addition, CFD analysis alone also allows for a comprehensive analysis of wire and tube heat exchanger running at certain operating conditions, and eventually, the heat exchanger capacity as well as the heat exchanger efficiency can be calculated based on the simulation results.

Acknowledgements

Financial support by the Center of Research and Development (*Lembaga Penelitian dan Pengabdian Kepada Masyarakat*) of Universitas Negeri Surabaya and the directorate of research and community development (*Direktorat Riset dan Pengabdian Kepada Masyarakat*) of Institut Teknologi Sepuluh Nopember is highly acknowledged. Authors also would like to thank to the heat transfer laboratory of Universitas Negeri Surabaya for their technical support.

Conflict of interest

The authors declare that there is no conflict of interest.

Nomenclature

A	area (m^2)
Cp	specific heat of hot and cold fluid (J/kg K)
C	heat exchanger capacity (W/kg)
D	diameter (m)
f	optimization factor
g	acceleration of gravity (9.8 m/s^2)
GP	geometrical parameter
Gr	Grashof number
h	heat transfer coefficient (W/m^2 K)
H	height of heat exchanger (m)
k	conduction heat transfer coefficient (W/m K)
L	length (m)
$\dot{m}$	mass flow rate, (kg/s)
Nu	Nusselt number
p	pitch distance (m)
Pr	Prandtl number
Q	heat transfer rate (W)
r	radius (m)
R	heat resistance (K/W)
Ra	Rayleigh number
s	spacing (m)
T	temperature, °C or K
W	heat exchanger width (m)
ε	heat exchanger effectiveness, n.d

Subscripts

el	refers to element of finite element model
f	refers to fluid
i,o	refers to inner and outer surface
in, out	refers to inlet flows and outlet flows

max	refers to maximum condition
s	refers to surface of the wire tube
t	refers to tube
w	refers to wire
x	refers to position at x direction
∞	refers to ambient/surrounding air

Greek symbols

α	thermal diffusivity (m^2/s)
β	thermal expansion coefficient (K^{-1})
θ	polar or cone angle measured from normal of surface, rad.
ρ_f	density of a fluid (kg/m^3)
σ	Stefan-Boltzmann constant
υ	kinetic viscosity (m^2/s)

Author details

I Made Arsana[1*] and Ruri Agung Wahyuono[2,3]

1 Faculty of Engineering, Universitas Negeri Surabaya, Department of Mechanical Engineering, Surabaya, East Java, Indonesia

2 Faculty of Industrial Technology and System Engineering, Department of Engineering Physics, Institut Teknologi Sepuluh Nopember, Surabaya, Indonesia

3 ENABLE Work Group, Laboratory of Advanced Functional Materials, Faculty of Industrial Technology and System Engineering, Department of Engineering Physics, Institut Teknologi Sepuluh Nopember, Surabaya, Indonesia

*Address all correspondence to: madearsana@unesa.ac.id

References

[1] Lum JMS, Clausing AM. An Investigation of the Air-Side Forced Convection Heat Transfer from Saw-Tooth Shaped, Multi-Layer, Wire-on-Tube Condensers. Reports. Urbana: Air Conditioning and Refrigeration Center, University of Illinois; 1997

[2] Hoke JL, Swofford TD, Clausing AM. An Experimental Investigation of the Air-Side Convective Heat Transfer Coefficient on Wire and Tube Refrigerator Condenser Coils. Reports. Urbana: Air Conditioning and Refrigeration Center, University of Illinois; 1995

[3] da Silva R, Melo EC. A Correlation for the air-side heat transfer coefficient of natural-draft wire-on-tube condensers. Articles in the Research Laboratories for Emerging Technologies in Cooling and Thermophysics. 2019. Available from: https://polo.ufsc.br/publicacoes/a-correlation-for-the-air-side-heat-transfer-coefficient-of-natural-draft-wire-on-tube-condensers.html

[4] Arsana IM, Wahyuono RA. Chapter: Nanofluid-Enhancing Shell and Tube Heat Exchanger Effectiveness with Modified Baffle Architecture. In: Miguel Araiz, editor. Heat Transfer-Design, Experimentation and Applications. UK: IntechOpen; pp. 95. DOI: 10.5772/intechopen.96996

[5] Arsana IM, Muhimmah LC, Nugroho G, Wahyuono RA. Enhanced heat transfer effectiveness using low concentration SiO_2-TiO_2 core-shell nanofluid in a water/ethylene glycol mixture. Journal of Engineering Physics and Thermophysics. 2021;**94**:423-430. DOI: 10.1007/s10891-021-02312-x

[6] Witzell OW, Fontaine WE. What are the heat transfer characteristics of wire and tube condensers? Refrigerating Engineering. 1957;**65**:33-37

[7] Witzell OW, Fontaine WE. Design of wire and tube condensers. Refrigerating Engineering. 1957;**65**:41-44

[8] Tagliafico L, Tanda G. Radiation and natural convection heat transfer from wire-and-tube heat exchangers in refrigeration appliances. International Journal of Refrigeration. 1997;**20**(7): 461-469

[9] Lee T-H, Yun J-Y, Lee J-S, Park J-J, Lee K-S. Determination of airside heat transfer coefficient on wire-on-tube type heat exchanger. International Journal of Heat and Mass Transfer. 2001;**44**:1767-1776

[10] Espíndola RS, Boeng J, Knabben FT, Hermes CJL. A new heat transfer correlation for natural draft wire-on-tube condensers for a broad geometry span. International Journal of Refrigeration. 2020;**114**:10-18. DOI: 10.1016/j.ijrefrig.2020.02.025

[11] Badwar SR, Joshi M, Mandale S, Kujur A. Experimentation and performance analysis of natural draft wire on tube condenser. International Journal of Recent Technology and Engineering. 2019;**8**(3):4303-4308. DOI: 10.35940/ijrte.C5174.098319

[12] Belman-Flores JM, Heredia-Aricapa Y, Garcia-Pabón JJ, Gallegos-Munoz A, Serrano-Arellano J, Pérez-Reguera CG. An approximate model of a multilayer wire-on-tube condenser operating with R134a and R600a: Experimental validation and parametric analysis. Case Studies in Thermal Engineering. 2021;**25**:100927. DOI: 10.1016/j.csite.2021.100927

[13] Nuntaphan A, Vithayasai S, Vorayos N, Vorayos N, Kiatsiriroat T. Use of oscillating heat pipe technique as extended surface in wire-on-tube heat exchange for heat transfer enhancement. International

Communications in Heat and Mass Transfer. 2010;**37**:287-292. DOI: 10.1016/j.icheatmasstransfer.2009.11.006

[14] Choi SH, Cho WH, Kim JW, Kim JS. A study on the development of the wire woven heat exchanger using small diameter tubes. Experimental Thermal and Fluid Science. 2004;**28**:153-158. DOI: 10.1016/80894-1777(03)00034-7

[15] Kushwah A, Pandey KD, Gupta A, Saxena G. Thermal modeling of heat exchanger: A review. International Journal of Advance Research in Engineering, Science and Technology. 2017;**4**(1):30-35

[16] Bansal PK, Chin TC. Modelling and optimisation of wire-and-tube condenser. International Journal of Refrigeration. 2003;**26**:601-613

[17] Gholap AK, Khan JA. Design and multi-objective optimization of heat exchangers for refrigerators. Applied Energy. 2007;**84**:1226-1239. DOI: 10.1016/j.apenergy.2007.02.014

[18] Quadir GA, Krishnan GM, Seetharamu KN. Modeling of wire-on-tube heat exchangers using finite element method. Finite Elements in Analysis and Design. 2002;**38**:417-434. DOI: 10.1016/S0168-874X(01)00079-8

[19] Imran AA, Jafal HM. Numerical modeling of wire and tube condenser used in domestic refrigerators. Journal of Engineering and Development. 2009;**13**(2):1-16

[20] Hadi SL, Arsana IM. Design and fabrication of woven matrix wire and tube heat exchanger [in Bahasa Indonesia]. Jurnal Rekayasa Mesin. 2014;**01**(03):56-63

[21] Putra TD, Arsana IM. Instrumentation system design for efficiency test in the wire and tube heat exchanger [in Bahasa Indonesia]. Jurnal Rekayasa Mesin. 2014;**01**(02):16-22

[22] Fikri MUK, Arsana IM. The effect of mass flow rate in the free convection wire and tube heat exchanger [in Bahasa Indonesia]. Jurnal Teknik Mesin. 2013;**01**(02):71-79

[23] Agung DP, Arsana IM. The effect inlet temperature in the free convection wire and tube heat exchanger [in Bahasa Indonesia]. Jurnal Teknik Mesin. 2013;**01**(02):80-85

[24] Arsana IM, Budhikardjono K, Susianto AA. Modelling of the single staggered wire and tube heat exchanger. International Journal of Applied Engineering Research. 2016;**11**(8):5591-5599. DOI: 10.1051/matecconf/20165801017

[25] Arsana IM, Susianto BK, Altway A. Optimization of the single staggered wire and tube heat exchanger. MATEC Web of Conferences. 2016;**58**:01017. DOI: 10.1051/matecconf/20165801017

[26] Akbar FR, Arsana IM. Effect of wire pitch on capacity of single staggered wire and tube heat exchanger using computational fluid dynamic simulation. International Journal of Engineering Transactions B: Applications. 2020;**33**(8):1637-1642. DOI: 10.5829/ije.2020.33.08b.22

[27] Arsana IM, Rahardjo MAH. Simulation study on efficiency of woven matrix wire and tube heat exchanger. International Journal of Engineering Transactions C: Aspects. 2020;**33**(12):2572-2577. DOI: 10.5829/ije.2020.33.12c.19

[28] Arsana IM. Empirical correlation of the free convection wire and tube heat exchanger efficiency [in Bahasa Indonesia]. Proceeding of National Postgraduate Seminar. 2002;01: C9-1-C9-5

Chapter 10

Multiscale Micro/Nanostructured Heat Spreaders for Thermal Management of Power Electronics

Huihe Qiu and Yinchuang Yang

Abstract

In this chapter, we describe surface modification techniques for enhancing heat/mass transfer and evaporation on heated surfaces. The effect of asymmetrical structure in designing a vapor chamber, patterned with multiscale micro/nanostructured surfaces will be introduced. The wettability patterned surface and its mechanism for improving the evaporation rate of a droplet and the thermal performance of nucleate boiling are discussed. An ultrathin vapor chamber based on a wettability patterned evaporator is introduced as a case for the application of the wettability pattern. Besides, modifying the surface with nanostructure to form a multiscale micro/nanostructured surface or superhydrophobic surface also enhances the phase change. Several types of heat spreaders are proposed to investigate the effects of multiscale micro/nanostructured surface and nanostructured superhydrophobic condenser on the thermal performance of the heat spreaders, respectively. The effects of multiscale micro/nanostructured evaporator surfaces with wettability patterns will be analyzed and experimental data will be presented.

Keywords: heat spreader, ultrathin vapor chamber, droplet evaporation, thin-film evaporation, multiscale wick, micro/nanostructure, wettability, wettability pattern, enhance heat transfer

1. Introduction

Two-phase heat spreaders, such as heat pipes, thermal ground planes, and vapor chambers, have been widely used in various areas owing to their high heat transfer efficiency. Improving the thermal performance of the heat spreaders becomes critical for the increasing demand for cooling requirements. Here, we introduce several surface modification techniques for enhancing the heat/mass transfer on heated surfaces and its application on the heat spreaders.

As shown in **Figure 1**, a type of two-phase spreaders, vapor chamber, is composed of wick structure, casing materials, working fluid and vapor core for vapor flow spreading. The evaporator is defined as the casing material and wick structure close to the heat source, while the condenser is defined as the casing material and wick structure close to the heat sink. The heated liquid working fluid in the evaporator vaporizes, spreading to the whole space of the vapor core. The vapor of the working fluid releases heat on the condenser side and condenses into liquid. Then it returns to the evaporator by capillary pressure through wick structure,

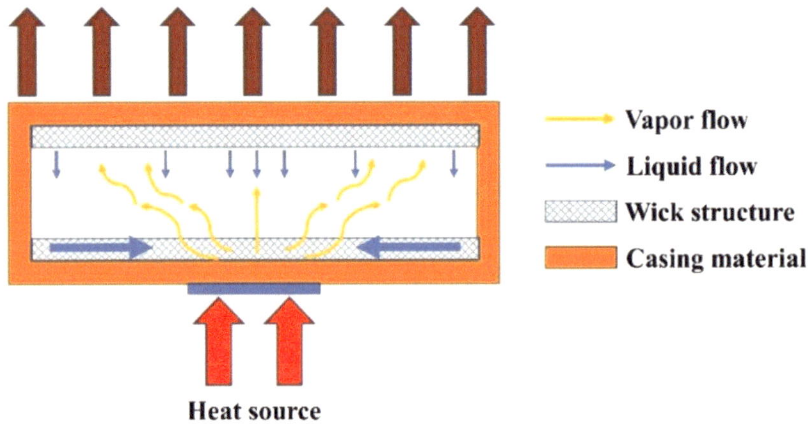

Figure 1.
Schematic of a vapor chamber structure and its working mechanism.

compensating the liquid working fluid of the evaporator. The heat pipe works similarly with the vapor chamber, but it only transfers heat in a one-dimensional way, while the vapor chamber transfers heat both in the horizontal and vertical directions.

Recently, lots of research has been conducted on the fabrication and thermal performance enhancement of the two-phase ultrathin heat spreader. Different casing materials, such as titanium [1], polymer [2, 3], and aluminum [4, 5], were used to develop the thin and light thermal ground plane. However, copper is still the most popular type of casing material for heat spreaders owing to its low cost, high thermal conductivity, and stable chemical characteristics. Thus, most investigations of different wick structures were conducted on the copper-based two-phase heat spreaders. Several types of wick structures were proposed to enhance the thermal performance of thin and ultrathin two-phase heat spreaders, including single arch-shaped sintered-grooved wick [6], bilateral arch-shaped sintered wick [7], mesh-grooved wick [6], pillar-mesh composite wick [8, 9], spiral woven mesh [10], copper fiber [11] and so on.

While many researchers focused on developing a new wick structure or modify wick structure in microscale to enhance the thermal performance of the two-phase ultrathin heat spreader, we tried to use some surface modification methods on the existing evaporator and the wick structure of the evaporator to change its wettability and structure in nanoscale for enhancing the thermal performance. The outline of the chapter is as follows. First, the mechanism of the wettability patterned surface on the nucleate boiling and the droplet evaporation are introduced, respectively. Then, an ultrathin vapor chamber with a wettability patterned evaporator is illustrated as an application case for the wettability patterned surface. After that, the fabrication of a micro/nanostructured surface is introduced and its effect on the wick structure and condenser surface is discussed. The experimental data of several types of heat spreaders with micro/nanostructured surfaces are analyzed.

2. The evaporation enhancement of a multicomponent droplet introduced by chemically wettability patterned surfaces

A molecular dynamic (MD) simulation study proved that the nanoscale wettability patterned surface can enhance the evaporation rate of water [12]. Here, the authors presented an experimental study to prove the enhancement of the evaporation of a multicomponent droplet also can be provided by a microscale wettability patterned surface.

2.1 Fabrication of the chemically wettability patterned surface

Chemically wettability patterned surface is composed of areas with different wettability. The authors used the method described below to fabricate the chemically wettability patterned surfaces.

A 2 cm × 2 cm prism transparent glass coated with a 100 nm thick Indium tin oxide (ITO) layer is selected as the substrate. ITO layer functions as a resistive heater and its two edges are deposited with 100 nm thick gold electrodes, which are used for wire bonding with a printed circuit board. After the deposition of gold electrodes, the sample is deposited with a 500 nm thick silicon dioxide as a passivation layer. Then a layer of Teflon, which is commonly used as a hydrophobic layer, is spun coated on the top of the silicon dioxide layer, and stabilized following the procedures suggested by DuPont™:

1. A layer of fluorosilane (1H, 1H, 2H, 2H-Perfluorodecyltriethoxysilane, Sigma-Aldrich Co. LLC.) is coated on the silicon dioxide layer to enhance the adhesion between Teflon and silicon dioxide layer. The sample is immersed in a 0.5% volume fraction of fluorosilane solution which used n-hexane as the solvent at least for 1 h. Then the sample is dried and stabilized in a 120°C oven for 1 h to finish the fluorosilane coating;

2. Then, the Teflon (DuPont AF 2401) is coated on the sample by spin-coating at a 1,000 rpm for 1 min;

3. After the spin coating process, the sample is dry in an oven with 40°C for 10 min to remove most of the solvent, 175°C for 10 min to remove 99% of the solvent remains, 250°C (above 5°C glass transition temperature) for 5 min to remove the last trace of the solvent and 330°C for 10 min to reach the maximum uniformity of coating thickness and to enhance the adhesion.

The Teflon layer was patterned to form a chemical wettability patterned surface with the following procedures as shown in **Figure 2**:

1. As the positive photoresist cannot attach to the Teflon, a 30 nm thick alumina layer is formed on the Teflon layer before photoresist coating by 300 cycles of Atomic Layer Deposition at 180°C using the Oxford OpAL ALD in Nanosystem Fabrication Facilities (NFF) in HKUST;

2. HPR506 positive photoresist is coated on the sample by spin coating with a 4,000 rpm for 30 s. Then the sample with a designed photomask is exposed to ultraviolet light with a 240 mJ/cm^2 energy after the sample is softbaked on a hot plate at 110°C for 1 min. The exposed sample is developed with FHD-5 for 1 min to reveal the pattern and hardbaked in a 120°C oven for 30 min to enhance the durability under dry and wet etching;

3. The unwanted alumina is removed by BCl_3/Cl_2 plasma in the Oxford Plasma Lab 80 Plus Reactive Ion Etcher. Then the unwanted Teflon is removed by O_2 plasma in the same machine;

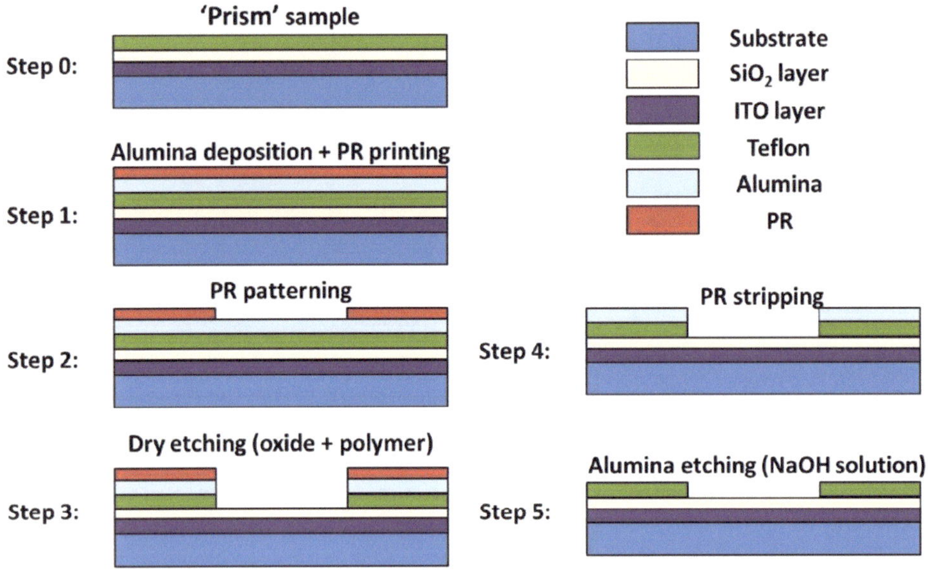

Figure 2.
Schematic fabrication procedures of a heterogeneous wetting surface [13].

4. Remain photoresist is removed by MS-2001 (Fujifilm Electronic Materials Co., Ltd.) for 2 min. The processed sample is checked with a microscope using fluorescence light to ensure no photoresist left;

5. Residual alumina is etched away by 5 mol/L NaOH solution for 10 min and the sample is checked with a wettability test.

After all the fabrication procedures are finished, the wettability patterned surface and a typical recording image during the droplet evaporation on the surface are shown in **Figure 3**. The wettability patterned surface is a hydrophilic surface (silicon dioxide on the prism glass) with square Teflon hydrophobic islands of D = 50 μm and pitch distance of L = 100 μm.

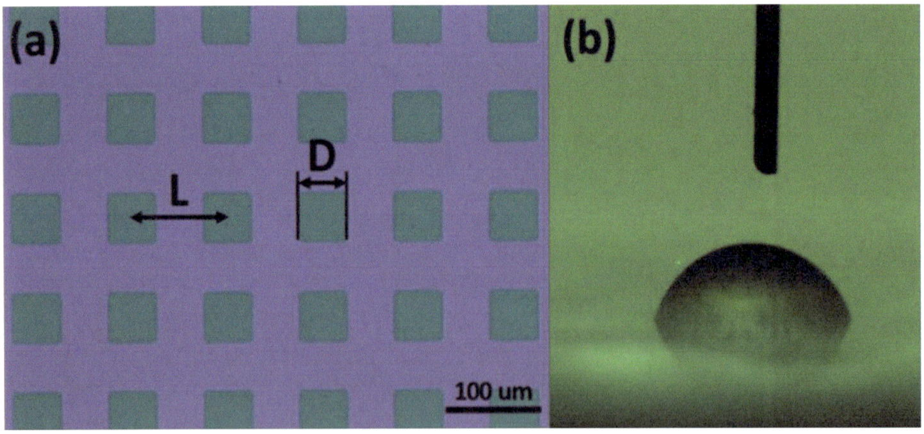

Figure 3.
(a) Microscopic image of a wettability patterned surface and (b) a typical recording image during the droplet evaporation on the surface [14].

The roughness of the sample is measured by a surface profiler (Dektak 150 Veeco). The roughness of the Teflon and the substrate is only 5 nm and 25 nm, respectively. Thus, its effect on evaporation is negligible. Finally, the sample is bonded on a Printed circuit board (PCB) for the experiments of multicomponent droplet evaporation with different sample temperatures.

2.2 Experimental setup and procedure of the droplet evaporation

The experimental setup is presented in **Figure 4(a)**. A Charge-coupled device (CCD) camera (MotionXtra HG-100 K, Redlake Co., Ltd.) with a long-distance zoom lens (Zoom 6000, Navitar Inc.) is adopted for recording the evaporation process to investigate the droplet evaporation characteristics including dynamic contact angle, base diameter, and evaporation duration of a droplet. Because the droplet profile is high symmetric, the focus plane of the camera is suitable to adjust to the cross-section of the droplet, as shown in **Figure 4(b)**. The droplet profile image is analyzed by a snake-based approach, which is shown in **Figure 4(c)**. Besides, the microscopic motion of the contact line is recorded by an inverted microscope (Eclipse TE2000-U, Nikon Instruments Inc.) connected to the CCD camera.

The droplet for the testing is generated by mixing deionized water and ethanol with an ethanol concentration varied from 0 vol. % to 15 vol. %. The droplets for each experiment are freshly generated before the test and the accuracy of the ethanol concentration is ±0.05 vol. %. The sample is heated and controlled at 40°C by a direct current (DC) power supply with a proportional-integral-derivative (PID) controller. The temperature difference between Teflon and the silicon dioxide layer

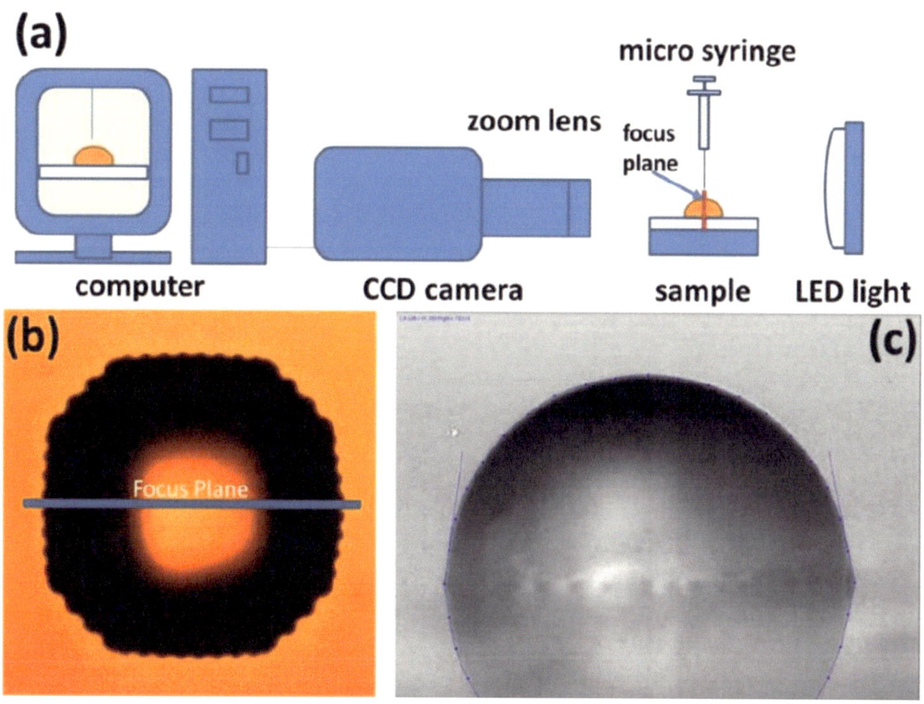

Figure 4.
Experimental setup and contact angle measurement for the droplet evaporation: (a) schematic of the experimental setup for the droplet evaporation; (b) bottom view of the droplet on a wettability patterned surface and corresponding focus plane of the contact angle measurement; and (c) the contact angle measurement result by the snake-based approach [14].

is negligible as the thickness of Teflon is only around 60 nm. Only if the sample is kept at 40°C for 5 min the test can be started.

The 1 μl ± 0.05 μl multicomponent droplet is produced by a Hamilton micro-syringe (Hamilton, 5 μl TLC syringe with a removable needle of gauge 33, 210 μm of outside diameter) and gently put on the heated homogenous or heterogeneous wettability surfaces. The location of the droplet is the same in every test to ensure the reproducibility of the experiments. After the droplet is put on the sample surface, the CCD camera starts to record the profile of the evaporation droplets or the microscopic motion of the contact line with the inverted microscope until the droplet vaporizes totally. To avoid the trace of the vaporized droplet affecting the next test, the sample is cleaned with acetone, ethanol, and deionized water, and dried with compressed air flow followed by a 105°C bake for 10 min in an oven. To ensure reproducibility, the location of the droplet is the same in every test and the whole experimental setup is kept at 23 ± 1°C and 50 ± 3% RH environment without any airflow. Besides, more than five times each experiment is conducted to identify reproducibility.

A hydrophilic surface, which is form by depositing only a layer of silicon dioxide on the sample, is chosen as a reference to investigate the enhancement effect because a homogenous hydrophilic surface always had a higher evaporation rate than that of a homogenous hydrophobic one [15].

2.3 Experimental result and discussion on the evaporation enhancement induced by a wettability patterned surface

First, the wettability of the multicomponent droplets on a hydrophilic (SiO2), hydrophobic (Teflon), and wettability patterned surfaces at 40°C are measured to confirm the wettability, as shown in **Table 1**.

Then, the evaporation duration of the droplets is recorded and shown in **Figure 5**. The evaporation duration of a droplet on a wettability patterned surface is around 10% shorter than that on a hydrophilic surface. Changing the ethanol concentration has an insignificant effect on the evaporation duration. The increase of the ethanol concentration leads to a shorter evaporation duration, which is contributed from the latent heat of ethanol is only 1/3 of water. In summary, the wettability patterned surface enhances the evaporation rate when compared with a hydrophilic surface.

The increase of the contact line length introduced by the wettability patterned surface plays a critical role in the enhancement. First, the mass flux of the evaporating flux ($j(x)$) of a slowly evaporating droplet on a hydrophilic surface can be calculated by [16]:

$$j(x) = j_0 \left[1 - \left(\frac{x}{r_s}\right)^2\right]^{-\frac{1}{2}+\theta/\pi} \tag{1}$$

Where, θ $(0 < \theta < \pi/2)$ is the instant contact angle during the evaporation, x is the length from the center of the drop which has a radius r_s. The factor of mass flux vapor (j_0) depends on the saturation pressure, vapor diffusivity, and far-field concentration. According to Eq. (1), the distribution of the evaporation flux for a droplet on a hydrophilic surface is non-uniform. The highest evaporation flux occurs at the contact line and it was also confirmed by Hu and Larson [17]. Besides, they derived an expression of the total evaporation rate of a droplet on a hydrophilic surface, which is:

$$\dot{m}(t) = -\pi RD(1-H)c_v\left(0.27\theta^2 + 1.30\right) \tag{2}$$

Ethanol concentration (vol.%)	Hydrophilic surface			Hydrophobic surface		Patterned surface
	Receding contact angle (°)	Advancing contact angle (°)	Initial contact angle (°)	Receding contact angle (°)	Advancing contact angle (°)	Initial contact angle (°)
0	62.1	88.7	78.6	84.5	134.9	82.1
5	54.8	82.5	75.0	78.3	130.5	78.5
10	45.7	75.8	69.3	67.5	127.2	73.4
15	33.2	71.2	59.5	61.1	124.1	69.0

Table 1.
Wettability data of the wettability patterned surface and the reference homogenous wettability surfaces under 40°C sample temperature [14].

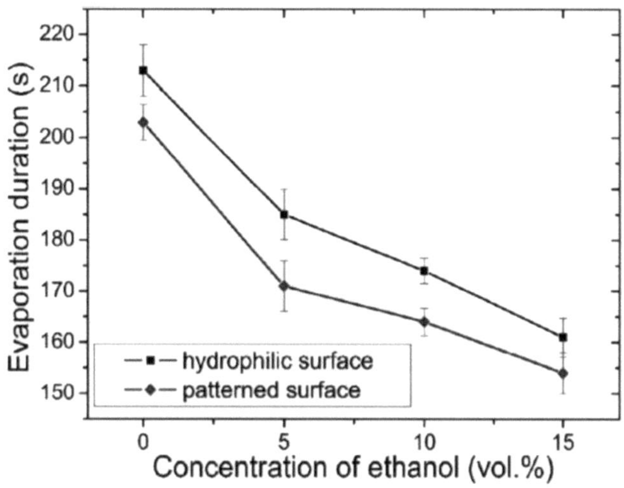

Figure 5.
Evaporation duration of a droplet with a range of ethanol concentration from 0 vol.% to 15 vol.% on homogenous (hydrophilic) and heterogeneous (wettability patterned) surfaces [14].

Where, R is the contact-line radius, D is the vapor diffusivity, and $(1-H)c_v$ is the vapor concentration difference. From Eq. (2) we can know the total evaporation rate is proportional to the perimeter of the contact line. Thus, the evaporation rate will be enhanced if the contact line is elongated. The wettability patterned surface increases the length of the contact line according to our observation, as shown in **Figure 6**. The spreading of the contact line is obstructed by the hydrophobic islands, leading to an elongation of the contact line along the islands' edge. Thus, although the contact area of a droplet on the wettability patterned surface is decreased, the length of the contact line is prolonged, resulting in faster evaporation of the droplet when compared with a homogenous hydrophilic surface under the same conditions. Moreover, a molecular dynamics simulation study also showed the increase of the total length of the hydrophobic island boundary in a nanoscale hydrophobic-hydrophilic pattern surface leads to an enhancement of the total evaporation rate [12].

Evaporation has been confirmed as the main phase-change mode in the evaporator of a two-phase heat spreader when the heat flux is less than 100 W/cm^2. Especially, thin-film evaporation which occurs nears the contact line contributes at

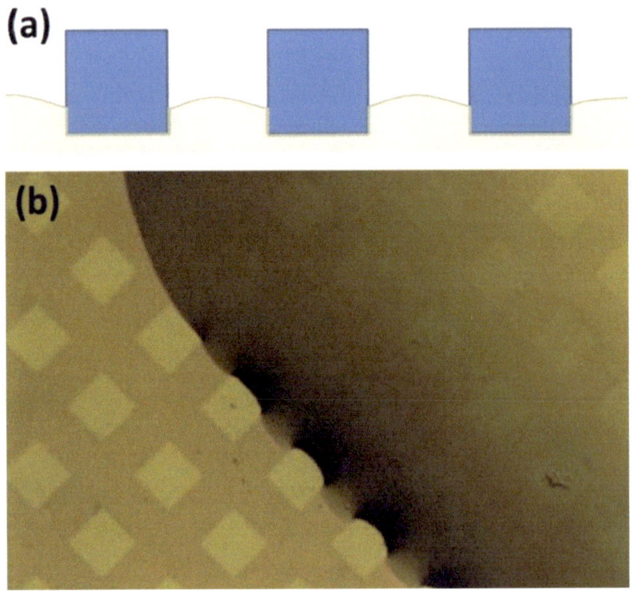

Figure 6.
The contact line profile of a multicomponent droplet (15 vol.% ethanol) on the wettability patterned surface: (a) the schematic of the contact line and (b) microscopy image [14].

least 50% even more than 80% of the total heat transfer in the evaporator [18–21]. Therefore, if the wettability patterned surface can be integrated into the evaporator of a two-phase heat spreader, the evaporation rate may be enhanced owing to the increase of thin-film evaporation are introduced by the elongated contact line. Then the thermal performance of the two-phase heat spreader can be enhanced. In the following section, an ultrathin vapor chamber with a wettability patterned surface integrated into the evaporator is presented. The effect of the wettability pattern on the thermal performance of the ultrathin vapor chamber will be discussed.

3. An asymmetric ultrathin vapor chamber with a wettability patterned surface on its evaporator

As shown in **Figure 7**, three types of ultrathin vapor chambers are designed and fabricated to investigate the effect of wettability pattern surface on the thermal performance of the ultrathin vapor chambers. The fabrication and the evaluation methods for the ultrathin vapor chamber are introduced. Multiscale micro/nanostructured wick structure is also adopted in the ultrathin vapor chamber and its effect will be also discussed in the next section.

3.1 Design and fabrication of the ultrathin vapor chambers with wettability patterned surface on its evaporator

Figure 7(a) presents the cross-section schematic of the ultrathin vapor chamber for case 1. Ultrathin copper (C1100P, 99.9%) is selected as the casing material and the #500 stainless-steel (SS 304) mesh covered with a layer of copper is adopted as the wick structure. A square array of micropillars integrated into the inner surface of the condenser is used for supporting the space of the vapor core. To investigate the effect of a multiscale micro/nanostructured surface, the only difference

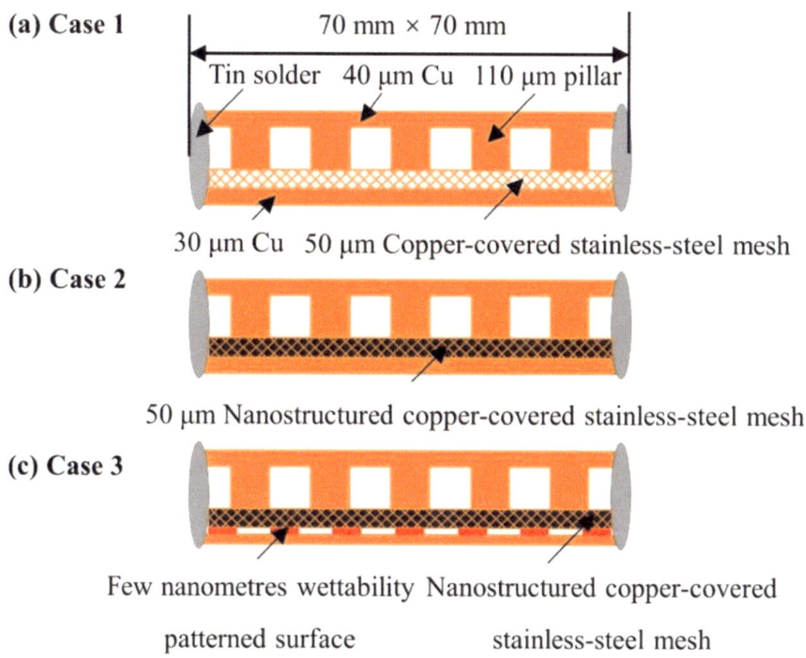

Figure 7.
The cross-section prototypes schematic of three ultrathin vapor chambers [22].

between case 1 and case 2 is the wick structure of case 2 is nanostructured to form a multiscale micro/nanostructured surface, as shown in **Figure 7(b)**. Case 3 which is shown in **Figure 7(c)** is developed based on the structure of case 2. A wettability pattern is fabricated on the inner surface of the evaporator casing material to examine its effect.

The square array of micropillars is fabricated through photolithography and electroplating, which is described below and shown in **Figure 8(a)**:

1. A 40 μm copper foil is cleaned with acetone, isopropanol, and DI water ultrasonic in sequence to remove organic contaminants. Then it is immersed into 20% sulfuric acid to dissolve the native oxide layer on its surface;

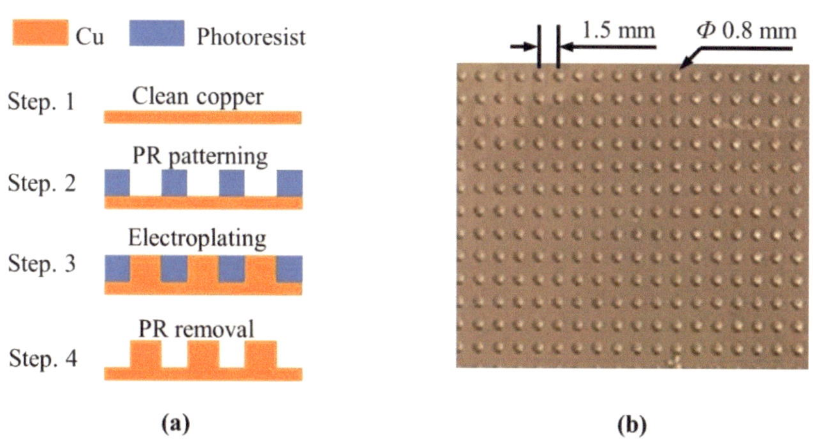

Figure 8.
(a) the fabrication procedures of the micropillar array and (b) the picture of micropillar array [22].

2. A positive photoresist AZ9260 is coated on the cleaned copper by spin-coating at 2,000 rpm for 30 s. Then the coated copper is softbaked on a 110°C hot plate for 3 min to remove most solvent in the photoresist. To ensure the exposure quality of the photoresist pattern mask, the copper should be put in an environment with a more than 60% RH for 10 min to do the rehydration process before the exposure. After that, the coated copper with a chromium photomask is exposed with 1200 mJ/cm^2 energy and then developed in an AZ400K developer diluted with 4 parts of water for 2 min. A microscope with fluorescent light is used to check the development quality. In the final of this step, a 120°C hardbaked for 30 min is adopted to enhance the adhesion between photoresist and copper for a long time electroplating process;

3. The photoresist patterned copper is immersed in an electroplating tank to do the electroplating process. The current is set at 0.07 A and the electroplating duration is around 15 h. When the electroplating process is completed, the micropillar shows a height of 110 μm and a diameter of 0.8 mm, as shown in **Figure 8(b)**. The pitch distance between two nearby micropillars is 1.5 mm;

4. The copper is immersed in MS2001 at 70°C for 5 min in Nanosystem Fabrication Facility-Hong Kong University of Science and Technology to remove the photoresist and the fabrication process is finished.

The wick structure of case 1, copper-covered #500 stainless-steel (SS 304) mesh (wire diameter: 25 μm, wire spacing: 25 μm) is fabricated with the following method. First, the mesh is activated at a current of 2 A for 2 min in a stainless-steel electroplating activation solution (From Beichen Limited Company, China) to enhance the bonding quality between the electroplated copper layer and the stainless-steel mesh. Then, the copper layer is formed by electroplating in a stationary solution with 0.8 M $CuSO_4$ and 1.5 M H_2SO_4 at a current of 1 A for 5 min, as shown in **Figure 9(a)**. The nanostructured copper-covered stainless-steel mesh of case 2 is fabricated by oxidizing the copper layer through a chemical surface modification method [23]. The method is firstly immersing the copper-covered stainless-steel mesh into an aqueous solution with 0.065 M K_2SO_8 and 2.5 M KOH at 70°C for 30 min. Then the mesh is rinsed with water and dried in a 180°C oven for 1 h. A flower-like nanostructure is grown on the surface of the copper layer, as shown in **Figure 9(b)**.

Figure 9.
(a) The stainless-steel mesh with a layer of copper and (b) the mesh with a layer of flower-like nanostructure [22].

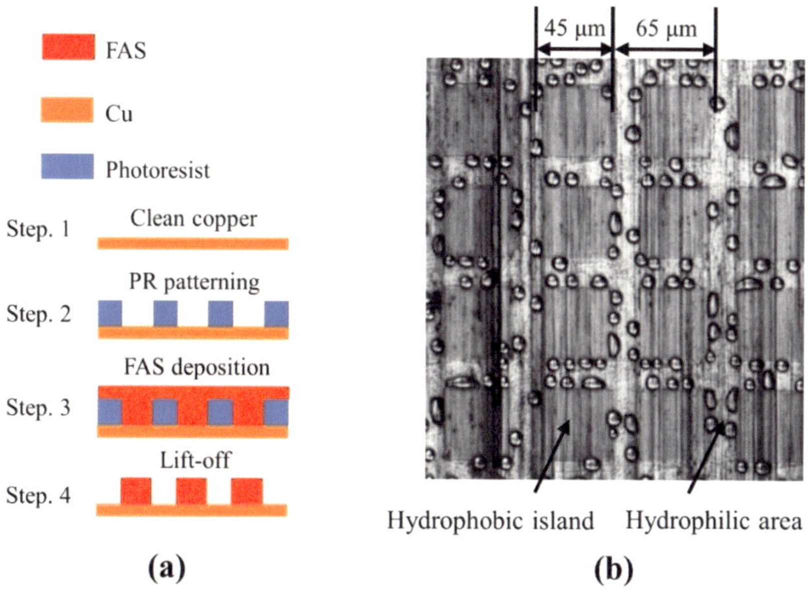

Figure 10.
(a) The fabrication procedures of a wettability pattern and (b) wettability check of a wettability patterned surface [22].

The wettability patterned surface for case 3 is fabricated with the following method, as shown in **Figure 10(a)**:

1. A 30 μm copper foil is firstly cleaned with the method described in the fabrication of the micropillar array;

2. A positive photoresist HPR506 is coated on the copper by spin-coating at 4,000 rpm for 30 s. Then the photoresist-coated copper is softbaked on a 110°C hot plate for 1 min. After that, the copper with a chromium photomask is exposed with 120 mJ/cm^2 energy and then developed in an FHD-5 develop for 1 min to reveal the photoresist pattern;

3. The photoresist patterned copper is immersed into a 1H, 1H, 2H, 2H-Perfluorodecyltriethoxysilane (FAS-17) solution and baked in a 120°C oven to form a hydrophobic layer which shows a 110° water contact angle and only a monolayer thickness [24]. The solvent of the Fluorosilane (FAS) solution is n-hexane, and the mass ratio of FAS is 1.6%;

4. The FAS-treated copper is immersed into an acetone solution to do the lift-off process. The parts with photoresist are dissolved into acetone. Thus, the FAS layer on the top of the photoresist is also removed. Then a wettability pattern is fabricated on the surface of copper.

Figure 10(b) shows a wettability check method using the condensation of water. The vapor condenses on the hydrophilic area, proving the success of the fabrication. The wettability pattern is composed of hydrophobic islands with a 45 μm side length and a 65 μm pitch distance between two nearby islands.

After the fabrication of all components, the mesh is sandwiched between the bottom and top casings, and then it is sealed by SnAg (97/3) solder. A tiny copper tube is connected to the inside of the chamber for the evacuating and feeding process. The dimensions of ultrathin vapor chambers are summarized in **Table 2**.

Item	Dimension
Bottom copper casing material (C1100P)	70 mm × 70 mm × 0.03 mm
#500 stainless-steel mesh (SS304)	60 mm × 60 mm × 0.05 mm
Micropillar array	70 m × 70 m × 0.1 mm
Micropillar diameter	0.8 mm
Pitch distance of the micropillars	1.5 mm
Top copper casing material (C1100P)	70 mm × 70 mm × 0.04 mm
Hydrophobic island side length (d)	45 μm
Pitch distance of the hydrophobic islands (P)	65 μm
The Thickness of the wettability pattern	1.34 nm

Table 2.
Dimensions of ultrathin vapor chambers [22].

3.2 Experimental setup of the thermal performance test for the ultrathin vapor chamber

When the assembly of the ultrathin vapor chamber is completed, a certain amount of water is filled into the inner space of the vapor chamber before the experimental tests. The vapor chamber is evacuated to 1 Pa and filled with different charge ratio η, which is defined as:

$$\eta = \frac{V_{water}}{V_{total}} \times 100\% \qquad (3)$$

Where, V_{water} is the volume of water inside the vapor chamber and V_{total} is the total volume of the inner space for the vapor chamber. As shown in **Figure 11(a)**, the evacuated ultrathin vapor chamber only has a less than 200 μm thickness.

The LW-9510 vapor chamber thermal performance measurement apparatus (LongWin Co., Ltd.) is adopted for evaluating the thermal performance of the vapor chamber. The vapor chamber is heated at the evaporator side by a heater with an 8 mm × 8 mm footprint and cooled on the outer surface of the condenser by an adjustable cooling fan with a 6.875 cfm flow rate at 22°C room temperature, as shown in **Figure 11(b)**. Besides, Teflon material is used as thermal insulation materials to reduce heat loss, making the maximum heat loss is only 0.41 W in the tests. One T-type thermocouple is attached to the center of the condenser outer surface (T_1) and 4 T-type thermocouples are attached at the side of the surface (T_2, T_3, T_4, and T_5), as shown in **Figure 11(c)**. A T-type thermocouple Five is inserted into the block below the center of the evaporator with a 1 mm distance to measure the heater temperature (T_h), which is shown in **Figure 11(d)**.

The authors examine the thermal performance of the ultrathin vapor chambers in the horizontal and vertical direction using horizontal thermal resistance (R_{hr}) and vertical thermal resistance (R_{vr}), respectively, which are defined as:

$$R_{hr} = \frac{T_1 - T_{avg(2-5)}}{Q} \qquad (4)$$

$$R_{vr} = \frac{T_b - T_{avg(1-5)}}{Q} \qquad (5)$$

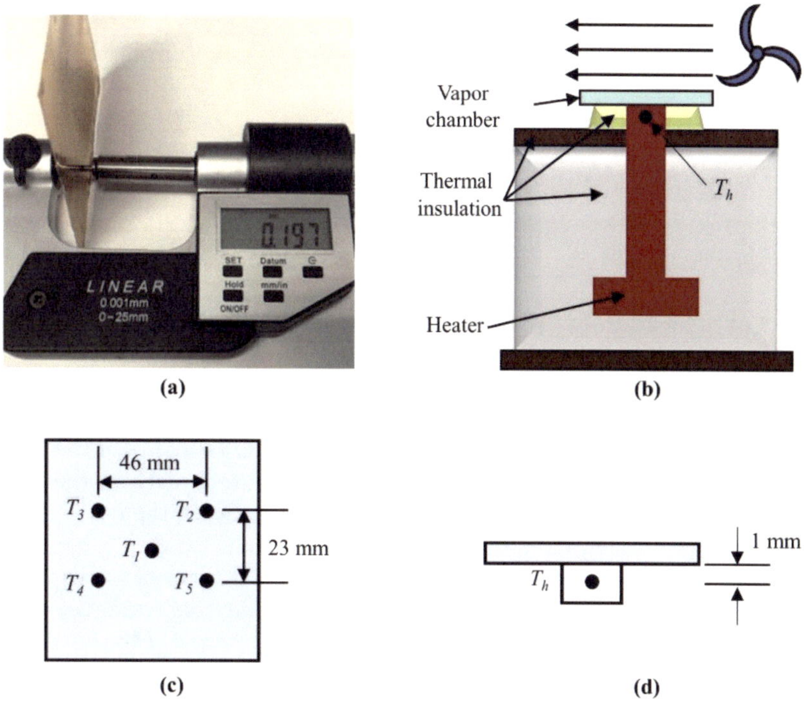

Figure 11.
(a) The thickness of an evacuated ultrathin vapor chamber and the schematic of the experimental setup: (b) apparatus; (c) temperature measurement points on the top of the vapor chamber; and (d) heater temperature measurement [22].

Where, Q is the input power measured by a power sensor of the LW9510, $T_{avg(1-5)} = (T_1 + T_2 + T_3 + T_4 + T_5)/5$ and $T_{avg(2-5)} = (T_2 + T_3 + T_4 + T_5)/4$. T_b is the temperature at the center of the evaporator outside surface, which is calculated based on Fourier's law:

$$T_b = T_h - ql / k_{Cu} \tag{6}$$

where l is the distance from the location of T_h to the center of the evaporator outside surface, which has a value of 0.001 m. k_{Cu} is the thermal conductivity of copper with a value of 401 W/(m·K). q is the input heat flux, which is defined as:

$$q = Q / A_h \tag{7}$$

Where, A_h is the footprint of the heater.

For characterizing the temperature uniformity of the vapor chamber and comparing it with copper, other materials, or commercial vapor chambers, the in-plane effective thermal conductivity (K_{eff}) is defined as:

$$K_{eff} = k_{Cu} R_{hr,Cu} / R_{hr,vc} \tag{8}$$

Where, $R_{hr,Cu}$ is the measured horizontal thermal resistance of copper.

The uncertainties are calculated based on the indirect parameter uncertainty analysis: if $F = F(x_1,,,x_2,,,\cdots,,,x_n)$, the uncertainty of F can be calculated by:

$$U_F = \left[\left(\frac{\partial F}{\partial x_1} U_1 \right)^2 + \left(\frac{\partial F}{\partial x_2} U_2 \right)^2 + \cdots + \left(\frac{\partial F}{\partial x_n} U_n \right)^2 \right]^{1/2} \tag{9}$$

Where U_1, U_2, ..., U_n are the uncertainties of x_1, x_2, ..., x_n, respectively. The maximum uncertainty of the heat flux q, horizontal thermal resistance R_{hr} and vertical thermal resistance R_{vr} under 95% confidence level are ±0.4%, ±6.1%, and ± 2.5%, respectively.

3.3 Effect of the wettability patterned surface on the thermal performance of the ultrathin vapor chamber

Figure 12 presents the experimental results of the thermal performance of the ultrathin vapor chambers. In this section, the thermal performance of the vapor chamber case 2 and case 3 are compared with each other to investigate the effect of wettability patterned surface as the only difference between them is whether a wettability pattern is integrated into the evaporator or not.

Figure 12(c) and **(e)** show the horizontal thermal resistance of the vapor chamber case 2 and case 3, respectively. The horizontal thermal resistance of case 2 for all the charge ratios first quickly decreases with the increase of the heat flux from 8.59 W/cm^2 to 14.53 W/cm^2. Then it gradually decreases when the heat flux increases from 14.53 W/cm^2 to 23.91 W/cm^2. On the other hand, the horizontal thermal resistance always decreases with the decrease of the charge ratio at a certain heat flux. Thus, the minimum horizontal thermal resistance exists at the 23.91 W/cm^2 heat flux of 60.4% charge ratio, which is 0.170°C/W. Compared with case 2, case 3 shows a similar trend but a lower horizontal thermal resistance under the same test conditions. For example, the horizontal thermal resistance at the 8.59 W/cm^2 of 60.4% charge ratio is 0.666°C/W for case 2 while it is only 0.295°C/W for case 3, which is only 44.3% of the previous. The minimum horizontal thermal resistance of case 2 is only 0.055°C/W occurring at 23.91 W/cm^2 heat flux of 60.4% charge ratio. The comparison proves integrating a wettability pattern to the inner surface of the evaporator of the ultrathin vapor chamber can greatly enhance the horizontal thermal performance of the ultrathin vapor chamber.

Figure 12(d) and **(f)** present the vertical thermal resistance of the vapor chamber case 2 and case 3, respectively. The vertical thermal resistance of case 2 for large charge ratio (65.2, 67.6, and 70.1%) gradually decreases with the increases of the heat flux, proving case 2 with large charge ratios does not reach the partial dry-out condition. When the charge ratio is less than 65.2%, case 2 starts partial dry-out at the heat flux of 23.91 W/cm^2, due to an insufficient amount of water returning to the evaporator. Though a small charge ratio gives a lower vertical thermal resistance if partial dry-out does not happen, the minimum thermal resistance for large charge ratios is close to that for the small charge ratio. This is because case 2 with a large charge ratio can work at a higher heat flux, leading to a further decrease of the vertical thermal resistance. The minimum vertical thermal resistance of case 2 is 1.37°C/W, happening at the heat flux of 19.84 W/cm^2 of 62.8% charge ratio. However, though the trend of the vertical thermal resistance of case 3 is similar to case 2, case 3 does not shows an improvement of the vertical thermal performance. When the charge ratio is small (60.4% and 62.8%), case 3 starts partial dry-out at a lower heat flux which is 19.84 W/cm^2. Besides, the vertical thermal resistance for all the charge ratios at a certain heat flux of case 3 is larger than that of case 2. The minimum thermal resistance for case 3 is just 1.54°C/W at 23.91 W/cm^2 heat flux of 70.1% charge ratio. Therefore, the evaporator integrated with the wettability

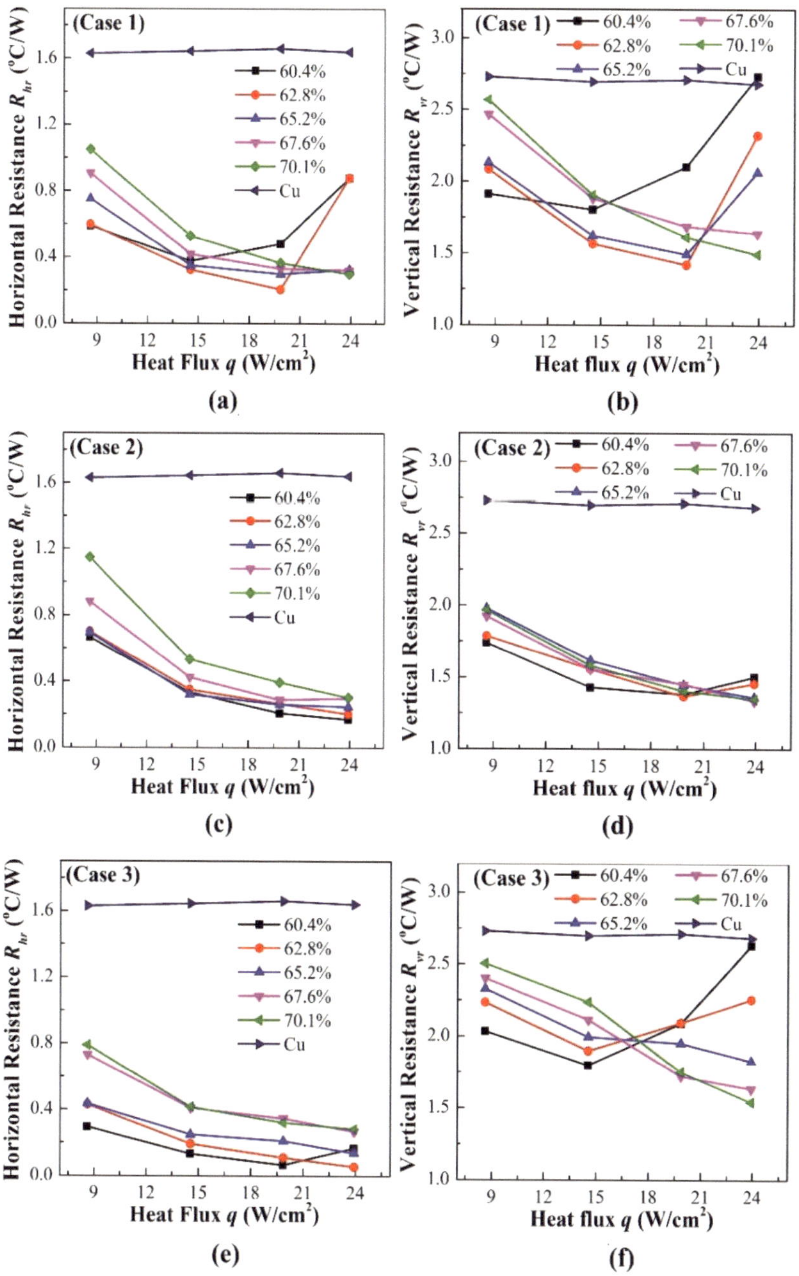

Figure 12.
Thermal performance of the ultrathin vapor chambers: Horizontal thermal resistance (a), (c), and (e); vertical thermal resistance (b), (d), and (f) [22].

pattern does not enhance the vertical thermal performance in this study. One possible reason is the large area of hydrophobic islands deteriorates the backflow ability of the wick structure. The capillary pressure determined the backflow ability of the wick structure, which can be calculated by [22]:

$$P_{Cap.} = \frac{2\sigma cos\theta}{r_{eff}} \tag{10}$$

where σ is the surface tension of the working fluid, r_{eff} is the effective radius of the wick structure and θ is the working fluid contact angle on the wick structure. Since part of the working fluid returns to the evaporator through the interface between the wick structure and the wettability patterned surface, the contact angle may be larger considered the effect of hydrophobic islands. Moreover, the area fraction of hydrophobic islands in this study reaches 47.9%, giving that $\cos\theta$ in Eq. (10) for case 3 may be much smaller than that for case 2. Thus, the backflow ability of the wick structure for case 3 is poorer than that of case 2, leading to lower cooling performance. Another possible reason is the photoresist, which may be left on the surface after the fabrication, releases non-condensable gas during the tests and deteriorate the backflow ability due to its hydrophobicity. Therefore, optimizing the design of the fabrication process may improve the vertical thermal performance of the ultrathin vapor chambers.

Generally, a wettability pattern can greatly enhance the temperature uniformity and is also promising to lower the vertical thermal resistance of a vapor chamber. In the next section, the authors will present that only changing the wettability of the condenser can also improve the temperature uniformity of a vapor chamber.

4. An asymmetric vapor chamber with a nanostructured superhydrophobic condenser

Compared with filmwise condensation occurring at a hydrophilic surface, dropwise condensation induced by a hydrophobic surface can largely increase the condensation heat transfer rate both under vacuum and atmosphere conditions [25]. Therefore, integrating a hydrophobic surface to the condenser of a vapor chamber is promising to enhance the thermal performance of the vapor chamber. In this section, the design and fabrication of a 70 mm × 70 mm × 3 mm asymmetric vapor chamber with a nanostructured superhydrophobic condenser are introduced firstly. Then the effect of the nanostructured superhydrophobic surface is discussed based on the experimental results.

4.1 Design, fabrication, and thermal performance test of the vapor chamber

Figure 13(a) illustrates the design of the vapor chamber. To compare the thermal performance with a conventional vapor chamber, the major difference between the proposed one and the conventional one is the wick-laid condenser for the conventional one is replaced by a nanostructured superhydrophobic condenser.

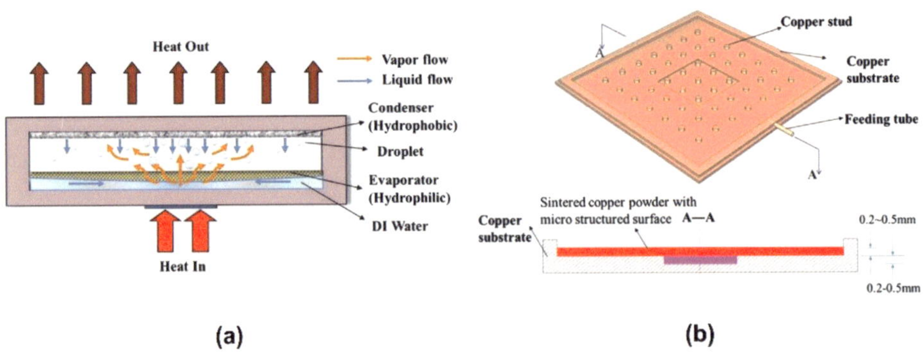

Figure 13.
(a) The schematic of the vapor chamber with a nanostructured superhydrophobic condenser and (b) the schematic of the evaporator base [26].

As shown in **Figure 13(b)**, the evaporator base with an area of 70 × 70 mm^2 is made of oxygen-free copper with a special configuration: (1) An array of studs is uniformly distributed on the copper substrate to prevent distortion caused by the pressure difference between the inner space of the vapor chamber and atmosphere; (2) A rectangular slot is set at the center of the substrate for sintering fine copper powders to enlarge capillary pressure in the center of the evaporator; and (3) A feeding tube is installed at the edge of the substrate to evacuate and feed working fluid. To fabricate the wick structure of the evaporator, a layer of size 57 μm copper powder is put into the center rectangular slot and then a layer of size 100 μm copper powder is covered the overall substrate. After that, a sintering process of the wick structure is performed in a 975°C hydrogen/nitrogen atmosphere for 2.5 h. This composite wick structure has an excellent backflow ability to prevent the occurrence of the partial dry-out as the smaller pores in the central wick provide higher capillary pressure.

The nanostructure of the condenser surface is fabricated with the method described in Section 3.1. Then, the nanostructured condenser is coated with a monolayer of FAS-17 using the method also described in Section 3.1. **Figure 14** shows the water contact angle of the condenser and the view of the flower-like nanostructure.

The thermal performance evaluation experimental setup is conducted on the LW-9510 platform which is described in Section 3.1. As the test power is much larger than that for the ultrathin vapor chamber, the cooling fan is replaced by a 170 × 80 mm^2 aluminum block attached to the outside surface of the condenser. The temperature of the aluminum block is set at 35°C and controlled by a water circulation system. Besides, the heater size is changed to 15 mm × 15 mm. In this study, the authors use the same parameter as that for the ultrathin vapor chamber, horizontal and vertical thermal resistance, to assess the thermal performance of this vapor chamber. The uncertainty of the heat load, temperature measurement, and the length measurement is ±0.1%, ±0.2%, and ± 0.4%, respectively [26].

4.2 The effect of the nanostructure condenser on the thermal performance of the vapor chamber

The thermal performance of the asymmetric vapor chamber is compared with that of a conventional vapor chamber and a copper with the same dimensions, as shown in **Figure 15**. The charge amount of the asymmetric vapor chamber is set at

Figure 14.
Features of the condenser: (a) front view of the condenser; (b) water contact angle of the condenser; and (c) amplified view of the nanostructure [26].

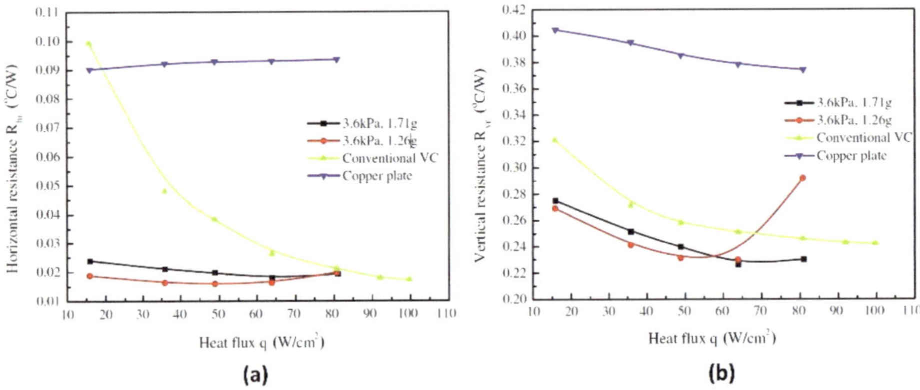

Figure 15.
Thermal performance comparison of the vapor chamber with a conventional vapor chamber and copper plate: (a) horizontal thermal resistance and (b) vertical thermal resistance [26].

1.71 g and 1.26 g. The pressure inside the vapor chamber is evacuated to 3.6 kPa. The conventional vapor chamber refers to a vapor chamber with symmetrical structure, which consists of two layers of copper mesh with 40 μm and 80 μm wire diameter sintered on both the evaporator and the condenser as a wick structure. Therefore, the evaporator and the condenser have the same hydrophilicity feature. The copper plate is a solid pure T2 copper solid heat spreader. The conventional vapor chamber and the copper plate have the same dimension as the asymmetric vapor chamber.

The asymmetric vapor chamber shows a much lower horizontal thermal resistance compared with the conventional vapor chamber and copper plate, as shown in **Figure 15(a)**. The horizontal thermal resistance of the asymmetric one is kept at around 0.02°C/W for all the testing heat flux. Especially, the horizontal thermal resistance at 15 W/cm^2 heat flux for the asymmetric one is only 1/5 of that for the conventional one which has a vertical thermal resistance even higher than that for the copper plate. The difference of the horizontal decreases with the increase of the heat flux and is eliminated at the heat flux of 82 W/cm^2. This proves that a nanostructured superhydrophobic condenser can enhance the temperature uniformity of the vapor chamber.

Figure 15(b) presents the vertical thermal resistance of different heat spreaders. Both types of the vapor chamber have a lower thermal resistance compared with the copper plate. Besides, the thermal resistance of the asymmetric one when the heat flux is less than 70 W/cm^2 is lower than that of the conventional one. However, when the heat flux is larger than 70 W/cm^2, the asymmetric one with 1.26 g water has a larger thermal resistance than that of the conventional one. This can be explained by insufficient water flowing back to the evaporator. Thus, increasing the amount of water may lower the vertical thermal resistance at the high heat flux for the asymmetric one, which is proved by the asymmetric one with 1.71 g shows a best vertical thermal performance at the heat flux of 82 W/cm^2.

In summary, the asymmetric vapor chamber with a nanostructured superhydrophobic condenser has a better thermal performance both in horizontal and vertical aspects when compared with the conventional vapor chamber. This can be explained as follows: the mode of the condensation on the condenser of the conventional vapor chamber is filmwise condensation which occurs at the hydrophilic surfaces while dropwise condensation is the mode for the condensation on the condenser of the asymmetric vapor chamber. Dropwise condensation has at least 3 times the heat transfer rate of filmwise condensation [25]. Thus, the thermal resistance of the condenser for the asymmetric vapor chamber is lower than that for the conventional one. Besides, droplets formed during dropwise condensation can return to the evaporator by directly contacting or falling into the evaporator surface, while liquid film formed during filmwise condensation only can return the evaporator through the wick structure. Therefore, dropwise condensation induced by the nanostructured superhydrophobic surface also improves the backflow ability of the wick structure. As a result, a larger amount of heat is removed and spread to the whole inner space in the asymmetric vapor chamber.

The authors presented wettability modification and patterning can enhance the thermal performance of vapor chambers in this and the last sections. In the next section, another surface modification method, which is growing nanostructured on the microstructure to form a multiscale micro/nanostructured wick structure, also can improve the thermal performance of vapor chambers.

5. Multiscale micro/nanostructured wick structure for two-phase heat spreaders

Sintering a porous microstructure on the inner surface of the vapor chamber to form a wick structure is the most common way in the fabrication of a vapor chamber. Researchers tried to use various kinds of porous microstructure to enhance the thermal performance of the vapor chamber, while little research was focused on growing a nanostructure on the porous microstructure to form a multiscale micro/nanostructured wick structure for enhancing the thermal performance before 2010. However, the nanostructure can enlarge the area of thin-film evaporation, increase the nucleation sites for boiling and improve the backflow ability of the wick structure without changing the microstructure [27, 28]. Thus, the authors investigated the effect of the multiscale micro/nanostructured wick structure on several types of two-phase heat spreaders.

5.1 Fabrication and the test of the multiscale micro/nanostructured wick structure

The wick structure of the asymmetric vapor chamber with a nanostructured superhydrophobic condenser, which is described in the previous section, is nanostructured by a chemical surface modification method presented in Section 3.1. The bare sintered wick structure shown in **Figure 16(a)** is coated with a layer of a thin-fin array of copper oxide, which is shown in **Figure 16(b)** and **(c)**. The thermal performance of the vapor chamber with a bare sintered wick structure is compared with that with a multiscale micro/nanostructured wick structure. The experimental setup is the same as described in Section 4.1. The uncertainties of vertical and horizontal thermal resistance for these vapor chambers are 0.7% and 0.8%, respectively.

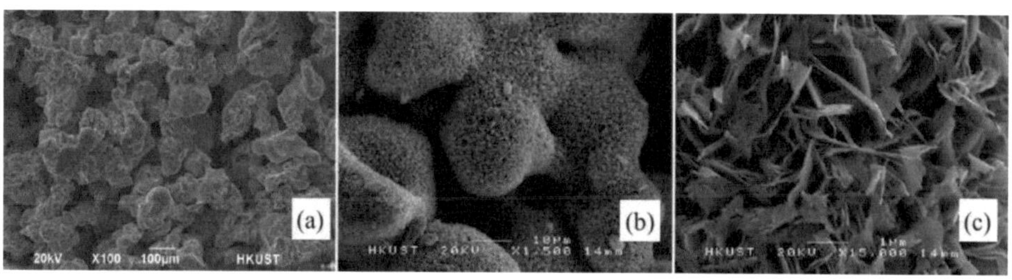

Figure 16.
Scanning electron microscope (SEM) images of the wick structures of the asymmetric vapor chamber with a nanostructured superhydrophobic condenser: (a) bare sintered powder wick structure; (b) low-magnification of multiscale micro/nanostructured wick structure; and (c) high-magnification of multiscale micro/nanostructured wick structure [29].

Moreover, the thermal performance of the ultrathin vapor chamber case 1, which only has a copper-covered stainless-steel mesh wick structure (shown in **Figure 9(a)**, is compared with that of vapor chamber case 2, which has a multiscale micro/nanostructured wick structure (shown in **Figure 9(b)**). The fabrications of other parts of these two types of vapor chambers can be found in sections 3.1 and 4.1. The parameters for assessing the thermal performance can be found in sections 3.2 and 4.1.

5.2 Effect of the multiscale micro/nanostructured wick structure on the thermal performance of two-phase heat spreaders

The thermal performance of two types of the asymmetric vapor chamber with different wick structures on the evaporator is compared with each other, as shown in **Figure 17**. The vertical thermal resistance for both vapor chambers decreases from around 0.63°C/W to 0.32°C/W when the heat flux increases from 5 W/cm^2 to the maximum test heat flux, which is shown in **Figure 17(a)**. However, the micro/nanostructured wick structure gives a quicker decrease of the vapor chamber at the low heat flux range compared with the bare sintered wick. For example, the vertical thermal resistance of the vapor chamber with micro/nanostructured wick shows a 0.25°C/W decrement with the increase of the heat flux from 5 W/cm^2 to 35 W/cm^2. While with the bare sintered wick, it only decreases 0.18°C/W. Besides, the micro/nanostructured wick structure enhances the vertical thermal performance for all the test heat flux compared with the bare sintered wick. These can be explained as follows: In the low heat flux range, thin-film evaporation is the dominant heat transfer mode for the evaporator. The thin-fin array nanostructure grown on the microstructure surface increase the superficial area of the wick structure, leading to an enlargement of the thin-film evaporation area. Thus, the heat transfer rate of the evaporator is improved. In the high heat flux range, nucleate boiling replaces the thin-film evaporation to become the dominant heat transfer mode in the evaporator. As shown in **Figure 16(b)**, the nanostructure roughens the wick structure when compared with the bare sintered wick shown in **Figure 16(a)**, resulting in more nucleation sites in the wick structure. Thus, the extent of boiling in the evaporator is enhanced and the heat transfer rate is increased. Moreover, the backflow ability of the multiscale micro/nanostructured wick structure is enhanced owing to its super hydrophilicity. Thus, the vapor chamber can work well at high heat flux conditions.

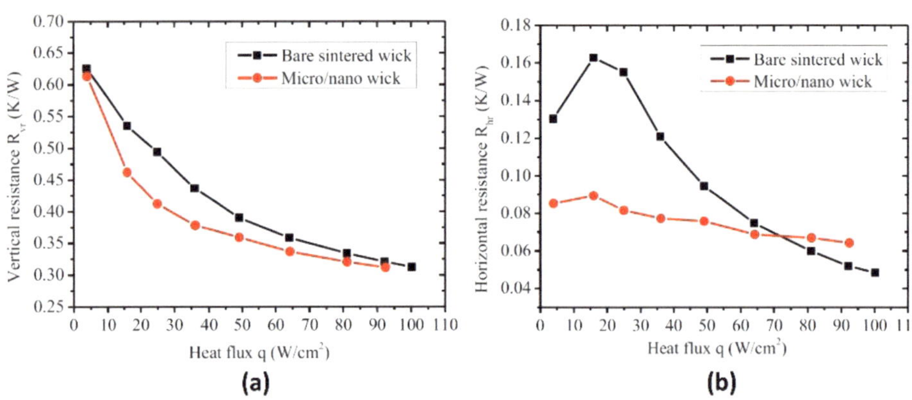

Figure 17.
Thermal performance comparison of the vapor chambers with different wick structures: (a) vertical thermal resistance and (b) horizontal thermal resistance [29].

The enhancement of the horizontal thermal performance, presented in **Figure 17(b)**, can be also explained with the advantages provided by the nanostructure. The improved thin-film evaporation at the low heat flux on the micro/nanostructured wick structure increases the amount of the vaporized liquid spreading to the whole space of the vapor chamber, resulting in a reduction of the horizontal thermal resistance by 30–50% within the heat flux ranging from 5 W/cm^2 to 35 W/cm^2. Besides, the micro/nanostructured wick structure gives a more stable horizontal thermal performance from low heat flux region to high heat flux region. Generally, the vapor chamber with a multiscale micro/nanostructured wick structure has a better horizontal thermal performance than that with a bare sintered wick structure.

The thermal performance of the ultrathin vapor chamber can be also enhanced by nanostructuring the wick structure to form a micro/nanostructured wick structure. As shown in **Figure 12(a)** and **(c)**, the difference of the horizontal thermal resistance between case 1 and case 2 under the optimum charge ratio (62.8% for case 1 and 60.4% for case 2) can be neglected within the range of heat flux from 8.59 W/cm^2 to 19.84 W/cm^2. However, it becomes obvious when the heat flux increases to 23.91 W/cm^2 owing to the occurrence of serious partial dry-out on the evaporator of case 2. A large increment of the vertical thermal resistance for case 1 is also found at the heat flux of 23.91 W/cm^2 while it for case 2 just slightly increases, which are shown in **Figure 12(b)** and **(d)**. These also prove the enhancement of the backflow ability induced by the multiscale wick. Besides, the vertical thermal resistance under the optimum charge ratio for case 2 is smaller than that for case 1 due to the enhancement of the thin-film evaporation that happened on the evaporator.

The authors also adopted the micro/nanostructured wick structure for another type of two-phase heat spreaders, the thermal ground plane, which works in a one-dimensional way [30]. By combining a two-layer structure with the multiscale wick, the thermal performance of the thermal ground plane is enhanced. Therefore, growing nanostructure on the microstructure surface to form a multiscale wick is an efficient way to improve the thermal performance of various kinds of two-phase heat spreaders.

6. Conclusion

In this chapter, several surface modification methods for improving the thermal performance of a two-phase heat spreader are introduced and discussed. Forming a wettability pattern on a surface enhances the evaporation of a droplet by elongating the length of the contact line. Thus, integrating the wettability pattern to the evaporator of a two-phase heat spreader is probably to enhance the thermal performance of the heat spreader, which is proved by the temperature uniformity enhancement of an ultrathin vapor chamber utilizing an evaporator with a wettability patterned surface. Changing the wettability of the condenser to superhydrophobic can also enhance the thermal performance by modifying the condensation mode from filmwise condensation to dropwise condensation which shows at least 3 times the heat transfer rate of the previous mode. Nanostructuring the microstructure to form a multiscale micro/nanostructured wick structure is also an efficient way to enhance the thermal performance of a two-phase heat spreader. It not only enhances the heat transfer rate on the evaporator but also increases the backflow ability of the wick structure, greatly improving the overall thermal performance of a two-phase heat spreader.

Acknowledgements

These works were supported by the Research Grants Council and Innovation Technology Commission (ITC) of the Government of Hong Kong Special Administrative Region (HKSAR) with RGC/GRF Project No. 617812, 16205018, 618210, ITS/530/09, and ITS/162/16FP, and the International Science and Technology Project of Huangpu District of Guangzhou City (2019GH02). We acknowledge assistance from the Nanosystem Fabrication Facility (NFF) of HKUST for the device/system fabrication.

Appendices and nomenclature

A_h	heater area
c_v	saturated water vapor concentration
d	hydrophobic island side length
D	vapor diffusivity
H	relative humidity
j	mass flux of evaporating flux
j_0	factor of mass flux vapor
k_{Cu}	thermal conductivity of copper
K_{eff}	in-plane effective thermal conductivity
l	distance from the heater to the center of the evaporator outside surface
m	evaporation rate
P	pitch distance of the hydrophobic islands
q	input heat flux
Q	input power
r_s	droplet radius
R	droplet contact line radius
R_{hr}	horizontal thermal resistance
R_{vr}	vertical thermal resistance
$R_{hr,Cu}$	measured horizontal thermal resistance of copper
$R_{hr,vc}$	measured horizontal thermal resistance of a vapor chamber
t	evaporation time from the beginning of the experiment
T	temperature
V_{water}	volume of water inside the ultrathin vapor chamber
V_{total}	total volume of the inner space for the ultrathin vapor chamber
x	length from a droplet center to the position
η	charge ratio of the ultrathin vapor chamber
θ	instant contact angle during the evaporation
π	Pi

Author details

Huihe Qiu* and Yinchuang Yang
The Hong Kong University of Science and Technology, Hong Kong SAR, China

*Address all correspondence to: meqiu@ust.hk

References

[1] Ding C, Soni G, Bozorgi P, Piorek BD, Meinhart CD, MacDonald NC. A flat heat pipe architecture based on nanostructured titania. Journal of Microelectromechanical Systems. 2010;**19**(4):878-184

[2] Oshman C, Shi B, Li C, Yang R, Lee YC, Peterson GP, et al. The development of polymer-based flat heat pipes. Journal of Microelectromechanical Systems. 2011;**20**(2):410-417

[3] Oshman C, Li Q, Liew L-A, Yang R, Bright VM, Lee YC. Flat flexible polymer heat pipes. Journal of Micromechanics and Microengineering. 2013;**23**(1):015001

[4] Chen Y-T, Kang S-W, Hung Y-H, Huang C-H, Chien K-C. Feasibility study of an aluminum vapor chamber with radial grooved and sintered powders wick structures. Applied Thermal Engineering. 2013;**51**(1-2):864-870

[5] Chang C, Han Z, He X, Wang Z, Ji Y. 3D printed aluminum flat heat pipes with micro grooves for efficient thermal management of high power LEDs. Scientific Reports. 2021;**11**(1):8255

[6] Li Y, Zhou W, He J, Yan Y, Li B, Zeng Z. Thermal performance of ultra-thin flattened heat pipes with composite wick structure. Applied Thermal Engineering. 2016;**102**:487-499

[7] Li Y, He J, He H, Yan Y, Zeng Z, Li B. Investigation of ultra-thin flattened heat pipes with sintered wick structure. Applied Thermal Engineering. 2015;**86**: 106-118

[8] Lewis R, Xu S, Liew L-A, Coolidge C, Yang R, Lee Y-C. Thin flexible thermal ground planes: Fabrication and scaling characterization. Journal of Micro-electromechanical Systems. 2015;**24**(6): 2040-2048

[9] Xu S, Lewis RJ, Liew L-A, Lee Y-C, Yang R. Development of ultra-thin thermal ground planes by using stainless-steel mesh as wicking structure. Journal of Microelectromechanical Systems. 2016;**25**(5):842-844

[10] Huang G, Liu W, Luo Y, Li Y. A novel ultra-thin vapor chamber for heat dissipation in ultra-thin portable electronic devices. Applied Thermal Engineering. 2019;**114726**

[11] Tang Y, Tang H, Li J, Zhang S, Zhuang B, Sun Y. Experimental investigation of capillary force in a novel sintered copper mesh wick for ultra-thin heat pipes. Applied Thermal Engineering. 2017;**115**:1020-1030

[12] Wan R, Wang C, Lei X, Zhou G, Fang H. Enhancement of water evaporation on solid surfaces with nanoscale hydrophobic-hydrophilic patterns. Physical Review Letters. 2015;**115**(19):195901

[13] He M. Evaporation of multi-component droplets on homogeneous and heterogeneous wettability surfaces [Thesis]. Hong Kong: The Hong Kong University of Science and Technology; 2017

[14] He M, Liao D, Qiu H. Multicomponent droplet evaporation on chemical micro-patterned surfaces. Scientific Reports. 2017;**7**:41897

[15] Birdi KS, Vu DT. Wettability and the evaporation rates of fluids from solid surfaces. Journal of Adhesion Science and Technology. 1993;**7**(6):485-493

[16] Ristenpart WD, Kim PG, Domingues C, Wan J, Stone HA. Influence of substrate conductivity on circulation reversal in evaporating drops. Physical Review Letters. 2007;**99**(23): 234502

[17] Hu H, Larson RG. Evaporation of a sessile droplet on a substrate.

The Journal of Physical Chemistry B. 2002;**106**(6):1334-1344

[18] Wang H, Garimella SV, Murthy JY. Characteristics of an evaporating thin film in a microchannel. International Journal of Heat and Mass Transfer. 2007;**50**(19-20):3933-3942

[19] Wang H, Garimella SV, Murthy JY. An analytical solution for the total heat transfer in the thin-film region of an evaporating meniscus. International Journal of Heat and Mass Transfer. 2008;**51**(25-26):6317-6322

[20] Ranjan R, Murthy JY, Garimella SV. Analysis of the wicking and thin-film evaporation characteristics of microstructures. Journal of Heat Transfer. 2009;**131**(10)

[21] Ranjan R, Murthy JY, Garimella SV. A microscale model for thin-film evaporation in capillary wick structures. International Journal of Heat and Mass Transfer. 2011;**54**(1-3):169-179

[22] Yang Y, Li J, Wang H, Liao D, Qiu H. Microstructured wettability pattern for enhancing thermal performance in an ultrathin vapor chamber. Case Studies in Thermal Engineering. 2021;**25**:100906

[23] Zhang X, Guo Y-G, Liu W-M, Hao J-C. CuO three-dimensional flowerlike nanostructures: Controlled synthesis and characterization. Journal of Applied Physics. 2008;**103**(11):114304

[24] Sugimura H, Hozumi A, Kameyama T, Takai O. Organosilane self-assembled monolayers formed at the vapour/solid interface. Surface and Interface Analysis. 2002;**34**(1):550-554

[25] Vemuri S, Kim KJ. An experimental and theoretical study on the concept of dropwise condensation. International Journal of Heat and Mass Transfer. 2006;**49**(3-4):649-657

[26] Sun Z, Chen X, Qiu H. Experimental investigation of a novel asymmetric heat spreader with nanostructure surfaces. Experimental Thermal and Fluid Science. 2014;**52**:197-204

[27] Weibel JA, Kim SS, Fisher TS, Garimella SV. Carbon nanotube coatings for enhanced capillary-fed boiling from porous microstructures. Nanoscale and Microscale Thermophysical Engineering. 2012;**16**(1):1-17

[28] Ranjan R, Garimella SV, Murthy JY, Yazawa K. Assessment of nanostructured capillary wicks for passive two-phase heat transport. Nanoscale and Microscale Thermophysical Engineering. 2011;**15**(3):179-194

[29] Sun Z, Qiu H. An asymmetrical vapor chamber with multiscale micro/nanostructured surfaces. International Communications in Heat and Mass Transfer. 2014;**58**:40-44

[30] Yang Y, Liao D, Wang H, Qu J, Li J, Qiu H. Development of ultrathin thermal ground plane with multiscale micro/nanostructured wicks. Case Studies in Thermal Engineering. 2020;**100738**